MW01493376

Advanced Basics of Geometric Measure Theory

Maria Roginskaya

CLÓ LOIGHIC/ LOGIC PRESS
Kilcock, Co. Kildare

LOGIC PRESS
The Maws, Kilcock
Co. Kildare, W23 D92N Ireland
Tel: +353-1-628-7343
e-mail: logicpress@gmail.com
www.logicpress.ie

ISBN (paperback): 978-1-326-36743-5
This book may be purchased through all booksellers, and is also available at a discounted price direct from www.lulu.com.

First published 2015

Maria Roginskaya is the sole author of this work.

Published by Logic Press
Printed and distributed by Lulu.com
3101 Hillsborough St., Raleigh, NC 27606-5436, USA

A book for everyone and no one
in honor of those who taught me

Contents

1   A natural measure   1

2   Axiom of choice, and other slippery matters   9

3   Sizes of sets   17

4   Borel sets: Explicit construction   23

5   Measure by a squeeze   29

6   A squeeze from one side   37

7   Three covering theorems   47

8   Measure and integral   55

9   Application of the covering theorems   63

10 Hausdorff dimension and sums, part I   71

11 Hausdorff dimension and sums, part II   77

Preface

These are lecture notes for a short course which I gave twice to Masters level students at the University of Gothenburg, and, in an abbreviated form, to a group of faculty members at the University of Lille. As many participants expressed their appreciation, I decided to publish the notes.

Every day the scope of mathematical knowledge grows. Combined with the current decrease in the average knowledge with which students arrive at university, this makes undergraduate study of Mathematics all the more rushed. It seems to the author that often the transition from basic Calculus to advanced Analysis is so abrupt that a student may fail to appreciate the difference. The main purpose of the book is to look closely at some results that are basic for modern Analysis and which fascinated the author when she was a student, and to show how they constitute a foundation for the branch of Analysis known as Geometric Measure Theory.

The secondary aim of the book is to give a straightforward but reasonably complete introduction to the definition of Hausdorff measure and Hausdorff dimension and to illustrate how non-trivial they can be. The course has no ambition to replace a serious course on Geometric Measure Theory, but rather to encourage the student to take such a course.

I address this book to students at late Bachelor or early Masters level, but it may also serve as easy (and hopefully pleasant) reading for a researcher in a different field of Mathematics. The only prerequisites are some experience of reading mathematical proofs, and preferably some knowledge of advanced algebra (for example, the definition of a group).

The original course had the format of double 45 minutes lectures and I have retained the indication of the best place to make a break in each lecture. This should facilitate an instructor who wishes to restructure it to single 45-50 minutes lectures. Some exercises form essential components of the course, and others are just asides, or need more than the basic prerequisites. The latter are marked with a star.

Maria Roginskaya. Gothenburg. June 2015.

Lecture 1
A natural measure

A natural measure would be a function which assigns a number to a set (and in this course we will suppose that this number is non-negative). It is natural to expect that if we add two sets their measures should be added as well, at least when no element is "counted twice", i.e. when the sets do not intersect each other. Two possible interpretation of this requirement are possible.

Definition 1.1. We say that a measure μ is *finitely additive* if for each (finite) collection of pairwise-disjoint sets $A_1, \ldots, A_n$ we have $\mu(A_1 \cup \ldots \cup A_n) = \mu(A_1) + \ldots + \mu(A_n)$.

Observation 1.2. *For each finitely additive measure μ we have $\mu(\emptyset) = 0$*

Proof. Take two sets $A_1 = \emptyset$ and $A_2 = \emptyset$. Those sets are disjoint, as $A_1 \cap A_2 = \emptyset$ (it sounds slightly odd that two sets, which are equal, are disjoint, but the empty set is often odd). As the measure is finitely additive

$$\mu(\emptyset) = \mu(A_1 \cup A_2) = \mu(A_1) + \mu(A_2) = \mu(\emptyset) + \mu(\emptyset).$$

The only number which satisfies the equation $x = x + x$ is 0. $\square$

Usually one requires from a measure even more than being finitely additive.

Definition 1.3. We say that a measure μ is *countably additive* if for each at most countable collection of pairwise disjoint sets $(A_j)_{j \in J}$ we have $\mu(\bigcup_{j \in J} A_j) = \sum_{j \in J} \mu(A_j)$.

Remark: As a finite set is less than countable, each countably additive measure is automatically finitely additive.

Another natural property for a measure of sets in a Euclidean space is that "non-deforming motions" of a set don't change its measure. To describe this property we first need to decide, which "motions" we are talking about. The natural motions of a Euclidean space are shifts (translations) and rotations. One can gather them all in one class, which is called the group of *isometries*.

Definition 1.4. A map $g : \mathbb{R}^n \mapsto \mathbb{R}^n$ is called an *isometry* if it doesn't change the distances between points of the space, i.e.

$$\text{dist}(x, y) = \text{dist}(g(x), g(y)).$$

Remark: One can talk about isometries on any metric space[1] (i.e. on each set provided with a "distance").

Exercise 1.1. Show that each isometry is a bijection.

Exercise* 1.2. Show that each orientation-preserving isometry of $\mathbb{R}^2$ is a translation, or rotation (around some point of the space). Is the same true for $\mathbb{R}^3$?

Exercise* 1.3. Show that each rotation and shift on $\mathbb{R}^2$ can be presented as a composition of two mirror reflections (and thus, in combination with the previous exercise, you show that each isometry of a plane is a composition of at most 3 reflections).

Exercise 1.4. One can introduce a metric on $2^{\mathbb{Z}}$ (i.e. the set of binary sequences) by letting $\text{dist}(x, y) = \sum_j \dfrac{|x_j - y_j|}{2^j}$. Verify that this is a metric, and suggest some isometries on the space.

Definition 1.5. We say that a measure μ is *invariant* with respect to a map ϕ if for each set K we have $\mu(K) = \mu(\phi^{-1}(K))$.

Exercise 1.5. Show that a measure μ is invariant with respect to a bijection ψ iff for each set K we have $\mu(K) = \mu(\psi(K))$.

---

[1] *A metric space* is a set M provided with a distance $d : M \times M \mapsto \mathbb{R}_+$, which satisfies the following conditions

1. $d(x, y) = 0$ implies $x = y$
2. $d(x, y) = d(y, x)$ (symmetry)
3. $d(x, z) \leq d(x, y) + d(y, z)$ (triangle inequality).

It would be most natural for a measure on a Euclidean space to be invariant with respect to each isometry of it. Yet, this is impossible[2] (unless every set has measure 0). We will prove a particular case of this statement.

Theorem 1.6. *There exists no measure on* $\mathbb{R}$, *which is countably additive, shift-invariant and has* $\mu([0,1]) = 1$.

Proof. Assume such a measure exists. Divide all numbers in the interval $[0,1]$ into equivalence classes, placing in the same class numbers x and y if the difference $x - y$ is a rational number. Consider a set E which contains exactly one number from each class. Let us number all rational numbers in the interval $[-1,1]$ (we get thus a sequence (r_j)). Denote now by E_j the set E shifted by the number r_j, i.e. $E_j = \{x + r_j : x \in E\}$. The construction implies that the countable union of disjoint sets $\cup E_j$ contains the interval $[0,1]$ and is contained in the interval $[-1,2]$. Now all sets in the union are shifts of the same set, and must thus have the same measure. This measure can not be 0, as the union has measure greater than the measure of the interval (i.e. 1). At the same time it can not be any positive number, as the union can be covered by three (shifted) copies of the interval, and thus should have measure at most 3. This gives us a contradiction, which proves the theorem. □

Exercise 1.6. Show that there is no countably additive measure invariant with respect to the rotations on the circle. (Hint: You can either imitate the proof above, or show that existence of the measure on the circle is equivalent to the existence of the measure on $\mathbb{R}$ by using the map $x \mapsto e^{i2\pi x}$. It is recommended that you do both.)

Exercise 1.7. Show that there is no countably additive measure in $2^{\mathbb{Z}}$ invariant with respect to all the isometries (see Exercise 1.4).

As we want to be able to measure sets and can not construct a countably additive measure on all the sets, we will, starting from the fourth lecture, construct a measure defined on only some of the sets (so called

---

[2]Assuming the axiom of choice, which will be closely discussed in the next lecture.

measurable sets). Yet, there is still a hope that we might be able to construct a measure, which is not countably but finitely additive (though those measures are much less interesting for any practical purpose). Indeed, such a measure can be constructed on $\mathbb{R}$ and on $\mathbb{R}^2$, but starting from $\mathbb{R}^3$ even finitely countable measures do not exist. For the proof, which comes after the break, we will use the following

Definition 1.7. We say that sets X and Y are *equi-decomposable* if it is possible to present them as disjoint unions $X = \bigcup_{j=1}^{n} X_j$ and $Y = \bigcup_{j=1}^{n} Y_j$ and find isometries ρ_j such that $X_j = \rho_j(Y_j)$.

Exercise 1.8. Show that the relation of being equi-decomposable is an equivalence[3] relation.

In order to show that on $\mathbb{R}^3$ a finitely additive measure invariant with respect to isometries does not exist, we will show (or rather give the main steps of the proof) that the unit ball of $\mathbb{R}^3$ is equi-decomposable with two copies of itself. This is usually called the Banach-Tarski Paradox.[4]

Break

We will start by showing that a sphere is equi-decomposable with two copies of itself[5]. This implies, by considering the rays from the center of the sphere, that a ball without the central point is equi-decomposable with two copies of itself.

---

[3] A relation $\sim$ is an equivalence relation if it is reflexive, symmetric, and transitive, i.e.

1. for each A we have $A \sim A$;

2. $A \sim B$ implies $B \sim A$;

3. $A \sim B$ and $B \sim C$ implies $A \sim C$.

[4] This is a weak form of the Banach-Tarski paradox. The strong form of Banach-Tarski paradox claims that any two bounded sets in $\mathbb{R}^3$ with non-empty interior are equi-decomposable.

[5] This is a result of Hausdorff.

To make the construction we will need the algebraic concept of a group.[6]

Exercise 1.9. Show that isometries form a group (with respect to composition).

Free groups

Let us consider a set $S = \{a_1, \ldots, a_n\}$, which we will later call the *generator* of the group. Take an alphabet $T = \{a_1, \ldots, a_n, a_1^{-1}, \ldots, a_n^{-1}\}$, and consider the set of all words with the alphabet T including the empty one, i.e. set of all finite ordered sequences of the the elements of T. It is clear that if we wish to consider those sequences as elements of a group we should be able to remove each instance where an letter stands next to its inverse. We will call this operation *concatenation* and the sequences in which there is no pair of elements inverse to each other standing next to each other, we will call *reduced*. Now consider the set G of all reduced words. We can introduce the multiplication $A * B$ on G by appending the word B after the word A, and making all possible concatenations until the sequence becomes reduced (as the only place for a concatenation to be possible is on the boundary between the words A and B, the operation described is uniquely defined). This group is called a *free group* and the size of the generator is called its rank.

Exercise 1.10. Verify that G is indeed a group (Notice: Here the most tricky part is to check associativity, so don't forget it.)

Exercise 1.11. Show that all free groups of the same rank are isomorphic, i.e. that there exists a bijection which preserves the group operation.

Exercise* 1.12. If you have worked with groups before, show that each group is a factor group of some free group (possibly with an infinite rank).

---

[6]A group is a structure consisting of a set and a binary group operation $*$ on it, which is associative, i.e. $(a * b) * c = a * (b * c)$; has a unit, i.e. an element for which $e * a = a * e = a$ for all elements a; and in which each element a has an inverse a^{-1}, for which $a * a^{-1} = a^{-1} * a = e$.

Exercise 1.13. Show that the group generated by the rotations

$$\phi = \begin{pmatrix} \frac{1}{3} & -\frac{2\sqrt{2}}{3} & 0 \\ \frac{2\sqrt{2}}{3} & \frac{1}{3} & 0 \\ 0 & 0 & 1 \end{pmatrix} \text{ and } \psi = \begin{pmatrix} 1 & 0 & 0 \\ 0 & \frac{1}{3} & -\frac{2\sqrt{2}}{3} \\ 0 & \frac{2\sqrt{2}}{3} & \frac{1}{3} \end{pmatrix}$$

is a free subgroup of rank 2. (Hint: Show by induction that each combination of ϕ, ψ and their inverses which ends with $\phi^{\pm 1}$ applied to $(1, 0, 0)$ is of form $(a, b\sqrt{2}, c)/3^k$, where a, b, c are integers, and b is not divisible by 3 unless the combination is trivial)

In general, one says that a group G *acts* on a set X, if the elements of G correspond to automorphisms[7] of X, the group operation of G corresponds to the composition of automorphisms, and the unit of the group corresponds to the identity map of the set. For simplicity we identify an element of the group with the automorphism to which it corresponds.

The core of the Hausdorff construction is the following.

Theorem 1.8. *If a group G is a free group of rank 2, and it acts on the set X, so that the elements of $G \setminus \{e\}$ have no fixed point, then it is possible to present X as a disjoint union $A_1 \cup A_2 \cup B_1 \cup B_2$, and find two elements $a, b \in G$, such that $X = A_1 \cup a(A_2)$ and $X = B_1 \cup b(B_2)$.*

Proof. For each point $x \in X$ consider its orbit $\mathcal{O}(x) = \{g(x) : g \in G\}$. As G is a group any two points have either disjoint or identical orbits (which gives us equivalence classes). Choosing one point x_0 from each orbit class, we can identify each point x of the orbit with an element of G, such that $g(x_0) = x$ and this identification is bijective, as there is no fixed point.

Let a and b generate the group G. For each orbit place the points which correspond to the reduced words in G which start from a in the set $\widetilde{A_1}$, those which starts with a^{-1} place in $\widetilde{A_2}$, the ones which starts from b place in $\widetilde{B_1}$ and the ones starting from b^{-1} place in $\widetilde{B_2}$. Observe that $X = \widetilde{A_1} \cup \widetilde{A_2} \cup \widetilde{B_1} \cup \widetilde{B_2} \cup \{e\}$, and the union is disjoint. Now, the second letter in a word starting with a^{-1} (if there is one) can be any of a^{-1}, b, b^{-1}

---

[7]An automorphism in this context is an bijective map of the set onto itself.

and thus $\widetilde{A_1} \cup a(\widetilde{A_2}) = X$, and similarly $B_1 \cup b(B_2) = X$, and the unions again are disjoint.

We have now two copies of X and also $\{e\}$ as a leftover. To deal with it let N be the set of all words which are sequences consisting only of a^{-1}. Notice that $N \subset \widetilde{A_2}$. Set $A_1 = \widetilde{A_1} \cup N \cup \{e\}$, and $A_2 = \widetilde{A_2} \setminus N$. Now, $X = A_1 \cup a(A_2)$, with the union being disjoint, and $X = A_1 \cup A_2 \cup B_1 \cup B_2$. Which proves the theorem. $\qquad\square$

Consider a free subgroup G of rank 2 of the rotations of the sphere. We can not apply the Theorem 1.8 directly, as there are fixed points to the elements of $G \setminus \{e\}$. But as each non-trivial rotation of the sphere has only two fixed points, there are only countably many points on the sphere which are fixed for some element of $G \setminus \{e\}$. As the union of the corresponding orbits is again countable, we see that one can apply the Theorem 1.8 to the remainder of the sphere.

Corollary 1.9. *There is a countable set D such that $S \setminus D$ is equi-decomposable with two copies of itself.*

In order to show that one can decompose a sphere in two copies of itself, it suffices to show that for each countable set D the sets S and $S \setminus D$ are equi-decomposable with respect to the group of rotations.

Theorem 1.10. *Let D be a countable set. Then $S \setminus D$ and S are equi-decomposable, i.e. there exists a disjoint pair of sets A and B, and a rotation ρ, such that $A \cup B = S \setminus D$ and $A \cup \rho(B) = S$, while $A \cap \rho(B) = \emptyset$.*

Proof. Let us choose a rotation axis which doesn't meet any point of D. Now for each n and each pair of points $x, x' \in D$ (not necessarily distinct) there are at most finitely many rotations ϕ with the chosen axis, such that $\phi(x) = x'$. Thus, there are at most countable many rotations with the chosen axis for which the orbits of the points of D are not either disjoint, or cyclic. As there are uncountably many rotations around each axis, we can thus pick a rotation ρ such that each point of D has infinite (non-cyclic) orbit, and all the orbits are disjoint.

Now, let $B = \rho^{-1}(D) \cup \rho^{-2}(D) \cup \ldots$. We see that $D \cap B = \emptyset$ and $\rho(B) = D \cup B$. Thus, if $A = S \setminus (D \cup B)$, we have $S \setminus D = A \cup B$ (disjoint union), and $S = A \cup \rho(B)$ (disjoint union). $\qquad\square$

7

Exercise 1.14. Show that a ball without one point is equi-decomposable to a whole ball with respect to the group of isometries (Hint: Use the ideas of the proof of the Theorem 1.10).

Exercise 1.15. Observe that Theorem1.10, Theorem 1.8, and exercise 1.14 do, together, imply the (weak) Banach-Tarski paradox.

Lecture 2
Axiom of choice, and other slippery matters

It is a generally accepted view in the modern mathematical community, that a mathematical theory consists of a (small) number of statements which are called *axioms*, and a (much greater) number of statements which can be derived from the axioms by applying a chain of certain logical rules (see [2]). Those latter statements are called *theorems, propositions, lemmas* and many other names, which mathematicians find appropriate. The main purpose of mathematical research is to provide the chains of arguments (called *proofs*), which would allow us to attach[1] new statements to a theory[2]. It is quite possible that adopting different systems of axioms, we would obtain theories, which include different (possibly opposite) statements. One calls a theory which includes two opposite statements simultaneously a *contradictory* theory. While a contradictory theory is a legitimate mathematical theory, it is not interesting, as it contains every statement.

It is, thus, very important for a mathematician to know a theory with which one chooses to work is not contradictory. As the theory is based

---

[1] There are several points of view on the matter: Some people assume that a statement becomes included in a theory only after somebody provides a proof, i.e. that a theory is *created* by the mathematicians; others believe that each statement is either true or false, and so the mathematicians *discover* it; and then there is all the range in between those two extremes. These are matters of personal philosophy, and none of these beliefs is wrong, but

it may be useful to keep in mind that the writer's opinion is close to the second (formalistic) point of view.

[2] In reality it is close to impossible to read a proof written in this way, pretty much as it is close to impossible to read a programme written in computer code. So the proofs we see in books and articles are compromises (as the programming languages are compromise between human and computer language). One is supposed to provide a collection of "clear enough" hints, which would allow any trained mathematician to write down a rigorous proof.

on the axioms, it is evident that the fewer axioms we accept the greater is the chance that the theory built on them is not contradictory. This is one of reasons why we try to use only a few simple statements as the set of axioms. (Even better would be to work with a set of axioms for which it would be known that the theory based on them is not contradictory, but unfortunately even for the simplest interesting sets of axioms this is shown to be impossible.) On the other hand, without enough axioms we cannot have a rich enough theory. So the number of axioms which one is willing to accept, as any other gambling decision, is a matter of personal preference. Nevertheless, most mathematicians accept the basic axioms of the set theory[3], which are necessary to avoid Russell's paradox[4] and still have the ability to work with non-finite sets.

Originally the most controversial axiom of the ZFC theory of sets is the Axiom of Choice, which is less simple than the rest of them.

Axiom of choice. For each collection of non-empty sets $\mathcal{F}$ there exists a function $g : \mathcal{F} \to \bigcup_{X \in \mathcal{F}} X$, such that $g(X) \in X$, for all $X \in \mathcal{F}$. That is for each collection of non-empty sets we may chose an element from each set.

It has been shown that if the ZFC without the Axiom of Choice is consistent, then adding the Axiom of Choice will not make the theory inconsistent. This and a large number of interesting applications has made the Axiom of Choice accepted by most mathematicians[5]. The following three important theorems are equivalent to the Axiom of Choice, subject to the rest of the ZFC axioms.

Definition 2.1. We say that a set S is *well-ordered* (by an order relation $\preceq$), if it is totally ordered (i.e. for any x, y in S, we have $x \preceq y$ or $y \preceq x$) and every non-empty subset of S has a least element.

Exercise 2.1. Show that every totally ordered set has at most one least element.

---

[3] It should be noticed that there exists several sets of axioms which define different theories of sets. The usually accepted one seems to be the Zermelo-Fraenkel set of axioms (ZFC), but at this point the writer starts to be out of her depth.

[4] The set which contains all the sets which do not contain themselves.

[5] The Axiom of Choice is accepted within this course

Theorem 2.2 (Zorn's lemma[6]). *If in a set S a relation of (partial) order $\preceq$ is such that each subset well-ordered by the relation $\preceq$ has an upper bound, then the set has an element x which is maximal with respect to the relation $\preceq$, i.e. for each y the relation $x \preceq y$ implies $x = y$.*

Theorem 2.3 (Well-ordering theorem). *On each set one can introduce a relation with respect to which the set is well ordered.*

Definition 2.4. A subset of a partially ordered set, which is totally ordered by the relation (i.e. each two elements are comparable) is called a *chain*.

Theorem 2.5 (Hausdorff's maximal principle). *In each partially ordered set there exists a maximal chain (under inclusion).*

We will show the equivalence by showing the sequence of implications: Axiom of choice $\Rightarrow$ Zorn's lemma $\Rightarrow$ Well-ordering theorem $\Rightarrow$ Hausdorff's maximal principle $\Rightarrow$ Axiom of choice.

Axiom of Choice implies Zorn's lemma. First, let a subset U contain all its upper bounds. Then, take one of those upper bounds x. If $x \preceq y$ for some y, then y is also an upper bound of U and thus $y \in U$. But this implies $y \preceq x$, so $y = x$. This means that x is a maximal element in S and the lemma is proven. So, we may assume that each subset has an upper bound which does not belong to the subset.

Consider the collection $\mathcal{F}$ of all well-ordered subsets of the set S (including an empty set). There is an obvious bijection from this collection on the collection of their upper bounds which does not belong to the set (remember that each element is an upper bound of an empty set). As the latter collection by the above assumption consists of non-empty sets, we can use the axiom of choice. Thus, we can fix a function $g : \mathcal{F} \to S$, such that for each well-ordered subset X, $g(X)$ is an upper bound of X, and $g(X) \notin X$. Now, let us consider a collection $\mathcal{G}$ of well-ordered sets which satisfy the property that for each x in $G \in \mathcal{G}$, $x = g(\{y \in G : y \prec x\})$.

---

[6]This is a modification of Zorn's lemma. In real Zorn's lemma one talks not about well-ordered, but totally-ordered subsets (i.e. subsets in which each two elements are comparable).

Every set in this collection contains as the minimal element $g(\emptyset)$ and is built in the same unique way (the only difference is when the set ends).

Definition 2.6. We call a subset B of a well-ordered set G *a segment*, if for each $z \in B$ and $y \in G$ the relation $y \prec z$ implies $y \in B$.

We show rigourously later, that for any two sets in $\mathcal{G}$ one of them is a segment of the other. This implies that $G^* = \bigcup_{G \in \mathcal{G}} G$ is also a well-ordered set, and satisfies $\forall x \in G^* : x = g(\{y \in G^* : y \prec x\})$.

Exercise* 2.2. Prove the last statement.

Consider $G' = G^* \cup \{g(G^*)\}$: This set should be in $\mathcal{G}$, as it is a well-ordered set, which satisfies the condition. On the other hand $G' \subset G^* = \bigcup_{G \in \mathcal{G}} G$ fails, and this gives us the contradiction, which completes the proof. It remains then show, that for any two sets of $\mathcal{G}$ one of them is a segment of the other.

Break

It follows immediately from the definition that a union of segments is a segment again, and a segment of a segment is a segment again. It is easy to see that for each well-ordered set G each $x \in G$ the set $\{y \in G : y \prec x\}$ is a segment. By considering the least element which does not belong to a segment we see that all segments of a well-ordered set has to be of this form, except for the whole set. We will denote the segment $\{y \in G : y \prec x\}$ as $G(x)$. In this notation the condition defining $\mathcal{G}$ becomes $x = g(G(x))$.

Exercise* 2.3. Every well-ordered set G either has a maximal element x and then $G = \{x\} \cup \bigcup_{y \in G} G(y)$, or it does not have a maximal element and then $G = \bigcup_{y \in G} G(y)$.

Exercise* 2.4. Show that the order on the elements of the well-ordered set G implies the order (by inclusion) on the corresponding segments, and

12

thus the collection of the segments of a well-ordered set is well-ordered by inclusion.

Now we consider two sets $G_1, G_2 \in \mathcal{G}$. As the collection of all segments of G_1 is well-ordered by inclusion, we can use the method of *transfinite induction*[7]: Consider the least segment B of G_1 which is not a segment of G_2. By the exercise 2.3 (as each segment of a well-ordered set is well-ordered) there are two possibilities: If $B = \bigcup_{x \in B} G_1(x)$, then as each smaller segment is a segment of G_2, so is their union, and B is a segment of G_2, which is a contradiction. Let now B have a maximal element x. As $G_1 \in \mathcal{G}$, $x = g(G_1(x))$. We know that $G_1(x)$ is a segment of G_2. Consider y - the least element of $G_2 \setminus G_1(x)$ (if the set is not empty). We see that $G_2(y) = G_1(x)$. As $G_2 \in \mathcal{G}$, $y = g(G_2(y)) = g(G_1(x)) = x$, and as $G_2(y) \cup \{y\}$ is a segment of G_2 this implies $B = G_1(x) \cup \{x\}$ is a segment of G_2, which is a contradiction. The last remaining possibility is that the set $G_2 \setminus G_1(x)$ is empty. This will be the case when G_2 is a segment of G_1. Thus, we have proven that either G_2 is a segment of G_1, or every segment of G_1 is a segment of G_2. As G_1 is a segment of itself, the later means that G_1 is a segment of G_2.

□

Exercise* 2.5. Show that for two well-ordered sets there exists a unique function which maps one of them on a segment of the other, preserving the order (which set to be mapped is also uniquely defined for the pair, unless the map is a bijection of the two sets). Hint: Use the transfinite induction.

The following proof is a typical way to use Zorn's lemma.

Zorn's lemma implies the Well-ordering Theorem. Consider a set X. We will show that there exists an order relation on it, which makes the set well-

---

[7]Transfinite induction is a method to use the order on a well-ordered, by considering the least element for which a statement fails. If we show that existence of such an element leads to a contradiction, then the statement holds for all elements of the set.

ordered[8]. Consider the set $\mathcal{R}$ of all pairs $(W, \prec)$, where $W \subset X$, and $\prec$ is a relation on it with respect to which the set W is well-ordered (if there is no such order on W, then W is not involved in any element of $\mathcal{R}$). Now introduce the following partial order relation on $\mathcal{R}$: $(W_1, \prec_1) \ll (W_2, \prec_2)$ if $W_1 \subset W_2$, $\prec_1 = \prec_2 \,|_{W_1}$, and W_1 is a segment of W_2.

Exercise* 2.6. Prove that if in a collection of well-ordered sets for each two one is a segment of the other then the union of all the sets in the collection is a well-ordered set, with respect to the unique order relation which agrees with all order relations on the sets. (You may have already done this before, but if not you can do it now.)

If some subset $\mathcal{W}$ of $\mathcal{R}$ is totally ordered by $\ll$, then the union

$$\bigcup_{(W, \prec) \in \mathcal{W}} W,$$

equipped with the correspondent order relation, is an upper bound of $\mathcal{W}$, and so we can apply Zorn's Lemma to $\mathcal{R}$ (one may notice that Zorn's lemma involves well-ordered subsets, but each well-ordered set is totally ordered, so we have an even stronger condition fulfilled).

Zorn's lemma gives as a maximal in $\mathcal{R}$ element $(W, \prec)$. If $W \subset X$ and $W \neq X$ then pick an element $x \in X \setminus W$, and consider $W' = W \cup \{x\}$ with the relation $\prec'$ which coincides with $\prec$ in comparison of elements other than x, and makes x greater than each other element.

Exercise* 2.7. Verify that $(W', \prec')$ is a well-ordered set and $(W, \prec) \ll (W', \prec')$.

As $(W, \prec)$ is a maximal element of $\mathcal{R}$, $(W, \prec) \ll (W' \prec')$ is a contradiction, which proves that $W = X$, i.e. $\prec$ is a relation, with respect to which X is well-ordered (it is quite clear that this relation is not unique). $\quad\square$

Well-ordering Theorem implies Hausdorff's maximal principle. Let X be a set partially ordered by a relation $\prec$. Introduce a well-ordering relation $\ll$ on it, and denote the least (with respect to $\ll$) element of X by m.

---

[8]While we have a lot of freedom to choose the relation, once upon a time it was very non-trivial to imagine an well-ordering relation on, for example, the set of real numbers

Let us define the following function on X

$$f(x) = \begin{cases} x, & \text{if } x \text{ is comparable by } \prec \text{ with all elements of } f(X_{\ll}(x)); \\ m, & \text{otherwise.} \end{cases}$$

Exercise* 2.8. Prove that this function exists. (Hint: Use transfinite induction.)

Exercise* 2.9. Prove that all elements of $f(X)$ are comparable by $\prec$.

If there is an element which is comparable (by $\prec$) with all elements of $f(X)$, then it belongs to $f(X)$, thus $f(X)$ is a chain with respect to $\prec$-order that is maximal under inclusion, among such chains. □

Hausdorff's maximal principle implies Axiom of Choice. Given $\mathcal{F}$ consider all pairs $(\mathcal{X}, f)$, such that $\mathcal{X} \subset \mathcal{F}$ and f is a function $f : \mathcal{X} \to \bigcup\limits_{X \in \mathcal{X}} X$, such that $f(X) \in X$. Introduce the order relation $(\mathcal{X}, f) \prec (\mathcal{Y}, g)$ if $\mathcal{X} \subset \mathcal{Y}$ and $f = g|_{\mathcal{X}}$. By Hausdorff's maximal principle we can find a maximal totally ordered subset Γ.

Exercise* 2.10. Show that on $\mathcal{Z} = \bigcup\limits_{(\mathcal{X}, f) \in \Gamma} \mathcal{X}$ a unique function g can be defined, such that $f = g|_{\mathcal{X}}$ for each $(\mathcal{X}, f) \in \Gamma$.

Exercise* 2.11. Show that for the above function g we have $g(X) \in X$.

It remains to show that $\mathcal{Z} = \mathcal{F}$. If not - pick a set $Y \in \mathcal{F} \setminus \mathcal{Z}$, an element $y \in Y$ and define

$$f(X) = \begin{cases} g(X), & X \in \mathcal{Z}; \\ y, & X = Y. \end{cases}$$

This function is defined on $\mathcal{Z} \cup \{Y\}$ and $(\mathcal{Z} \cup \{Y\}, f)$ is comparable with each pair in Γ, which contradicts the fact that Γ is maximal by inclusion. The contradiction proves that $\mathcal{Z} = \mathcal{F}$. □

15

Lecture 3
Sizes of sets

Ordinal numbers

Definition 3.1. A set is called an *ordinal number* or just an *ordinal* if its elements are sets which are also its subsets and which are well-ordered by the relation $\in$ of being an element.

Examples of such a set are $\emptyset, \{\emptyset\}, \{\emptyset, \{\emptyset\}\}, \{\emptyset, \{\emptyset\}, \{\emptyset, \{\emptyset\}\}\} \ldots$.

A reformulation of the definition is that $(X, \prec)$ is an ordinal iff $x = \{y \in X : y \prec x\}$ for every $x \in X$.

Exercise 3.1. Show that an ordinal is also well-ordered by inclusion (replacing belonging by inclusion in the definition gives an equivalent one).

Notice, that if a set X is well-ordered by inclusion, and every element $x \in X$ is also a subset $x \subset X$, then transfinite induction shows that $x = \{y \in X : y \prec x\}$ for every $x \in X$, i.e. that the set X is an ordinal. Indeed, let x is the first element such that $x \neq \{y \in X : y \prec x\}$. Then if $y^* = \max\{y \prec x\}$ exists, we know that $x \supset y^* = \{y \prec y^*\}$. But, as $x \neq y^*$ there is at least one element $x^* \in x \setminus y^*$. If $x \prec x^*$, then $x^* \in x \subset x^*$ i.e. $x^* \in x^*$, which contradicts to the axiom of regularity. By the same reason $x \notin x$. So the only element which can be in $x \setminus y^*$ is y^*. Thus, $x = y^* \cup \{y^*\} = \{y \prec x\}$. On the other hand if the maximal element in $\{y \prec x\}$ doesn't exist, then

$$\{y \prec x\} = \bigcup_{y^* \prec x} \{y \prec y^*\} = \bigcup_{y^* \prec x} y^* \subset x,$$

as $y^* \prec x \Leftrightarrow y^* \subset x$ and for $y^* \prec x$ we have $y^* = \{y \prec y^*\}$. At the same time if $x \prec x^*$ or $x = x^*$, then $x^* \notin x$, by the same argument as in the previous case. So, also in the second case $x = \{y \prec x\}$, which exhaust the possibilities. This contradiction completes the induction.

Exercise 3.2. Show that existence of an order preserving bijection for well-ordered sets is an equivalence relation.

Exercise 3.3. Show that every well-ordered set is order-isomorphic (has a bijection, which preserves the order) to a unique ordinal.

This last statement allows us to reintroduce the natural relation $\preceq$ on well-ordered sets, and provides us with a natural representative for each equivalence class (with respect to equivalence $x \approx y = (x \preceq y) \wedge (y \preceq x)$). Notice that $\preceq / \approx$ is a total-order relation.

Cardinal numbers

It is more common to talk about the *cardinality* of a set.

Definition 3.2. We say that two sets have the same *cardinality* if there exists a bijection between them.

Note: The difference from the equivalence for ordinals is that we don't need any relations on the sets.

Exercise 3.4. Show that having the same cardinality is an equivalence relation.

Definition 3.3. An abstract representative of the equivalence class of the sets of the same cardinality as X is denoted $\mathrm{Card}(X)$ and called a cardinal number of X.

Notice that for any two sets, well-ordering them by Well-Ordering Theorem, we will automatically have a bijection of one of them onto a subset of another. This seems like a good candidate for an order relation on sets, but as we remember the relation which well-orders a set is not unique.

Theorem 3.4. *If there exists a bijection f from a set A onto a subset of a set B, and also a bijection g from the set B onto a subset of A, then there exists a bijection between A and B.*

Proof. Without loss of generality we may assume that $A \cap B = \emptyset$. Introduce the following notations: $n(x)$ for $x \in A \cup B$ is the maximal amount of f^{-1}, g^{-1}, which can be applied to x (sometimes it is ∞). Now let $A_k = \{x \in A : n(x) = k$, $B_k = \{x \in B : n(x) = k$, $k = 0, 1, \ldots, \infty$.

18

Observe that $f(A_k) = B_{k+1}, g(B_k) = A_{k+1}$ and $f(A_\infty) = B_\infty$. Now define

$$h(x) = \begin{cases} f(x), & x \in A_{2k}; \\ g^{-1}(x), & x \in A_{2k+1}; \\ f(x), & x \in A_\infty. \end{cases}$$

It is easy to verify that the image of h is $B_\infty \cup \bigcup\limits_{j=0}^{\infty} B_j = B$, and that the map is an injection. But then h is the required bijection between A and B. □

The theorem shows that there is natural order on the cardinal numbers. Indeed, as before, we say that $A \leq B$ when there exists a bijection from A onto a subset of B. Then $A \sim B$ if $(A \leq B) \wedge (B \leq A)$, and the relation $\leq / \sim$ is a total order relation on the cardinal numbers.

The simplest way to show that there are plenty of different cardinal numbers is to show the following.

Theorem 3.5. *For each set X we have* $\mathrm{Card}(X) \neq \mathrm{Card}(2^X)$, *where 2^X denotes the set of all subsets of X.*

Proof. Assume there exists a bijection $g : X \to 2^X$. Consider $S = \{x \in X : x \notin g(x)\}$. As g is a bijection we may look at $s = g^{-1}(S)$. Now, if $s \in S$, then $s \notin S$, and, on the other hand, if $s \notin S$ then $s \in S$ - both are contradictions, which proves the theorem. □

Exercise* 3.5. Prove the theorem above in the case $X = \mathbb{N}$. (The idea is just to rewrite the proof, and trace how it works.)

Definition 3.6. The cardinality of the set of integers is *countable.*

Exercise 3.6. Show that

1. $\mathrm{Card}(\mathbb{N}) = \mathrm{Card}(\mathbb{N} \cup \{p\})$, i.e that a countable set stays countable if we add an element to it;

2. A union of two countable sets is countable;

3. A union of countably many finite sets is countable;

4. A union of countably many countable sets is countable.

(Hint: Show the bijection explicitly.)

Break

Theorem 3.7. $\mathrm{Card}([0,1]) = \mathrm{Card}(2^{\mathbb{N}})$.

Proof. We will conduct the proof in two steps: First we will show that there exists a countable set A, such that there is a bijection between $2^{\mathbb{N}}$ and $[0,1] \cup A$. Then we will show that for a countable A we have $\mathrm{Card}([0,1]) = \mathrm{Card}([0,1] \cup A)$.

To make the first step of the proof we identify the set $2^{\mathbb{N}}$ with the set of sequences $(a_1, a_2 \ldots)$, where $a_j \in \{0,1\}$ (this is a canonical bijection where we assign to a set E the sequence (a_j) with $a_j = 1$ if $j \in E$ and 0 otherwise).

Now we consider the function $\phi((a_j)) = \sum_{j=1}^{\infty} \dfrac{a_j}{2^j}$. As $a_j \in \{0,1\}$ it is clear that $\phi((a_j)) \in [0,1]$. By considering binary representation of a number, we see that the function is surjective. Let us analyze the sequences on which ϕ is not injective, i.e. $\sum_{j=1}^{\infty} \dfrac{a_j}{2^j} = \sum_{j=1}^{\infty} \dfrac{b_j}{2^j}$. Let k be the least index for which $a_k \neq b_k$. We see that, $\sum_{j=k}^{\infty} \dfrac{a_j}{2^j} = \sum_{j=k}^{\infty} \dfrac{b_j}{2^j}$. Without loss of generality we may assume that $a_k = 0$ and $b_k = 1$. Then, $\sum_{j=k}^{\infty} \dfrac{a_j}{2^j} \leq \dfrac{1}{2^k} \leq \sum_{j=k}^{\infty} \dfrac{b_j}{2^j}$, and the equality holds only if $a_{k+1} = a_{k+2} = \ldots = 1$, and $b_{k+1} = b_{k+2} = \ldots = 0$. Thus, this is the only case when $\phi(x) = \phi(y)$ and $y \neq x$. (Notice that we have also shown that at most two points can have

20

the same image.) Consider now the function[1]

$$\tilde{\phi}(x) = \begin{cases} \phi(x), & x = \min(\phi^{-1}(\phi(x))) \\ x, & \text{otherwise} \end{cases}$$

This function is injective, and surjective on $[0,1] \cup A$. The analysis above shows that for each k there are only finitely many pairs (x, y) such that $\phi(x) = \phi(y)$ and x and y differs first in the position k, and so the set A is no more than countable, which is what we wanted to show in the first step.

To do the second step of the proof, just pick an arbitrary countable sequence $B \subset [0, 1]$, and observe that $\text{Card}(B) = \text{Card}(B \cup A)$. $\quad\square$

It is clear that the natural order on ordinals is not the same as the order of their cardinal numbers, i.e. if an ordinal is less than another in the sense of ordinal inclusion, it can still have the same cardinal number. (On the other hand, if two well-ordered sets are equivalent to the same ordinal, then they necessarily have the same cardinality.)

Recall that all ordinal numbers are well-ordered by inclusion. The first ordinal, which has infinite[2] cardinality has by definition cardinality $\aleph_0$, and is called countable, as the set of integers has this cardinality. The first ordinal which has cardinality greater than $\aleph_0$, has cardinality $\aleph_1$, etc.

Continuum hypothesis

The set $2^{\aleph_0}$ (this set has the same cardinality as $\mathbb{R}$) has cardinality $\aleph_1$.

The Continuum hypothesis is proven to be undecidable in ZFC, i.e. neither it nor its negation will give a contradictory theory if one adds them to the axiom set (if ZFC is consistent itself). There are some theorems

---

[1]There is a natural order on $2^{\mathbb{N}}$: $(a_j) < (b_j)$ if $a_j = b_j$ for $j < k$ and $a_k < b_k$. This order is called *alphabetical*.

[2]There are several ways to define an infinite set - one can for example say that a set is finite if it has a different cardinality from each of its proper subsets.

proven under the assumption of the Continuum hypothesis, but it is not widely accepted as an axiom.[3]

Theorem 3.8. *There is no countably additive measure defined on all subsets of $\aleph_1$, such that $\mu(\aleph_1) = 1$ and $\mu(\{p\}) = 0$ for each $p \in \aleph_1$. (Hence on the assumption of CH there is no such measure on $\mathbb{R}$.)*

Proof. Let X be the least ordinal of cardinality $\aleph_1$ with respect to the order $\prec$. Using the bijection between X and $\aleph_1$ we may assume that the measure μ is defined on X. Consider a proper segment $X(y)$, where $y \in X$. By definition it is countable, and thus we can find (and fix) a bijection $f_y : X(y) \to \mathbb{N}$. Let $F(n,x) = \{y \in X : f_y(x) = n\}$. Observe that for each fixed n the sets $F(n,x)$ (parameterized by x) are disjoint.

Exercise 3.7. Verify the last statement.

As the measure is additive and $\mu(X) = 1$, for a fixed n there are no more than k sets $F(n,x)$ of measure greater than $1/k$ (or $\mu(X) > 1$). Considering the collection of sets of measure greater than $1/k$ for all positive integers k, we see that there are at most countably many sets $F(n,x)$, which have positive measure (i.e. whose measure does not vanish).

Thus, going through all n, there are only countably many pairs (n,x) for which the measure of $F(n,x)$ is positive. As X is by definition uncountable, there exists x, such that all $F(n,x)$ have measure zero. Now, $\bigcup_n F(n,x) = \{y : y > x\}$. We see that $X = \{z \le x\} \cup \bigcup_n F(n,x)$, where the first set is countable, and the last union is a countable union of sets of measure zero. This implies $\mu(X) = 0$, and so the existence of a non-trivial countably additive measure would lead to a contradiction. $\square$

---

[3]The mathematicians who proved the undecidability - Gödel and Cohen - were of the opinion that the Continuum Hypothesis should not be accepted (Gödel, who was a Platonist and believed all statements to be true or false, believed it is false). The reason for such a position is that acceptance of Continuum Hypothesis reduces the number of sets we can work with.

Lecture 4
Borel sets: Explicit construction

By now we should have come to the conclusion, that if we want to construct an interesting measure on an interesting set, it won't measure all the subsets. The question arises, which subsets we should measure.

Definition 4.1. A collection $\mathfrak{A}$ of subsets of X is called a σ-algebra if

1. $X \in \mathfrak{A}$;

2. If $E \in \mathfrak{A}$, then $(X \setminus E) \in \mathfrak{A}$;

3. For each collection $\{E_j\}_{j=1}^{\infty} \subset \mathfrak{A}$, we have $(\cup E_j) \in \mathfrak{A}$.

One says that a σ-algebra $\mathfrak{A}$ is generated by a collection of sets $\mathfrak{E}$ if it is the minimal σ-algebra, such that $\mathfrak{E} \subset \mathfrak{A}$.

Exercise 4.1. Show that intersection of any number of σ-algebras is again a σ-algebra. Derive from this that the minimal σ-algebra used in the definition above makes sense.

Exercise 4.2. Show that a σ-algebra is closed with respect to a countable intersection (i.e. if $\{A_j\}_{j=1}^{\infty}$ is a collection of sets from the σ-algebra, then $\bigcap_j A_j$ is also in the σ-algebra).

Here and everywhere else in this book the ball of radius r and center x is denoted $B(x, r) = \{y : d(x, y) < r\}$.

Definition 4.2. In a metric space a set E is called *open*, if for each $x \in E$ there exists $\epsilon > 0$, such that $B(x, \epsilon) \subset E$. A *closed* set is a set which has open complement.

If we consider the real line, it is natural to expect that at least open sets should be measurable.

Definition 4.3. The σ-algebra generated by open sets is called *the Borel σ-algebra* $\mathfrak{B}$ or just the (family of) *Borel sets*.

At this point we may try to construct a measure. But we leave it to the next lecture and for now will instead try to present the Borel sets in a more explicit way.

G_δ and F_σ

Given a collection of sets H we will denote by H_δ the collection of all countable intersections and by H_σ the collection of all countable unions of the sets from H. As $\bigcap\limits_{j=1}^{\infty} A = A$ and $\bigcup\limits_{j=1}^{\infty} A = A$, we see that $H \subset H_\delta$ and $H \subset H_\sigma$.

It is common in topology to denote in this context the collection of open sets by G, and the collection of all closed sets by F (thus G_δ is the collection of sets which can be presented as a countable intersection of open sets, and F_σ is the collection of all sets which can be presented as a countable union of closed sets).

Exercise 4.3. Show that for the real line (or for each metric space) each closed set is of type G_δ, and each open set is of type F_σ.

It is clear from the definition and Exercise 4.2 that each G_δ and F_σ set is a Borel set. It happens that this hints at a way to an explicit construction of Borel sets. Let us, though, first discuss why we do not say that G_δ or F_σ contains all the Borel sets. If $E = \cup_j(\cap_k G_{j,k})$, then $E = \bigcap\limits_{(k_1,k_2,...)} \cup_j G_{j,k_j}$, where the intersection is taken over all possible sequences of integers. While $\cup_j G_{j,k_j}$ are open sets, the intersection is taken over a non-countable set, which means it is not necessarily in G_δ.

Exercise 4.4. Verify that $\cup_j(\cap_k G_{j,k}) = \bigcap\limits_{(k_1,k_2,...)} \cup_j G_{j,k_j}$.

Exercise 4.5. Do the argument above for F_δ.

To keep track of the construction we will need the ordinal numbers. Let us recall that the ordinal numbers are well-ordered by set inclusion, and every element of an ordinal is also its subset. Recall also that the ordinal numbers themselves as a collection of sets are well-ordered by inclusion. It will be useful if we think a bit more about the structure of an ordinal number.

24

- The ordinal numbers are of two types, the ones which have a maximal element (in which case the maximal element is unique), and the ones which do not have a maximal element. We will call the ones with a maximal element *closed*, and the ones without a maximal element *open*. An open ordinal number is the union of all the ordinal numbers preceding it. A closed ordinal number is $o(x) \cup \{x\}$, where x is the maximal element of the ordinal, and $o(x)$ is an ordinal number, which we will denote $o-$.

- If o is an ordinal number, then there exists the least ordinal number, which is greater than o. This ordinal number is $o \cup \{o\}$. We will denote this number as $o+$.

- The union of any number of ordinal numbers is an ordinal number.

Proposition 4.4. *Set $G_{\emptyset} = G$. Let o be an ordinal. If o is closed and has a maximal element x, then define $G_o = (G_{o(x)})_{\delta\sigma}$; if o is open, then define $G_o = \bigcup_{\bar{o} \in o} G_{\bar{o}}$. Given this definition, the collection of Borel sets is the union $\mathfrak{G} = \bigcup G_o$, where o runs through all the ordinals[1].*

Proof. By the transfinite induction in o, we see that each set in $\mathfrak{G}$ is Borel. As $G = G_{\emptyset}$, and $F \subset G_{\delta} \subset G_{\emptyset+} \subset \mathfrak{G}$, we see that $\mathfrak{G}$ contains open and closed sets. This means we just have to check that $\mathfrak{G}$ is a σ-algebra.

$X \in \mathfrak{G}$ as X is open.

Let there exist $E \in \mathfrak{G}$, such that $X \setminus E \notin \mathfrak{G}$. Consider the least ordinal number o, such that G_o contains such a set E. $G_{\emptyset} = G$, so if $E \in G_{\emptyset}$ then $X \setminus E$ is closed and so is in $\mathfrak{G}$. Thus o is not the least ordinal number.

If o is open, than there exists even smaller ordinal $\bar{o}$ such that $E \in G_{\bar{o}}$. So o has to be closed, i.e. o has a maximal element.

By the definition of G_o, we can represent $E = \cup_j (\cap_k E_{j,k})$, where $E_{j,k} \in G_{o-}$. $X \setminus E = \cap_j (\cup_k (X \setminus E_{j,k}))$. Since we know that each set $X \setminus E_{j,k}$ is an element of $\mathfrak{G}$, we can find $o_{j,k}$, such that $(X \setminus E_{j,k}) \in G_{o_{j,k}}$. Let $o^* = \bigcup_{j,k} o_{j,k}$. Then $(X \setminus E_{j,k}) \in G_{o^*}$ for all j, k. This means that

---

[1] Talking about the set of all ordinals places us on a shaky ground, but the next theorem shows that everything is okey

25

$\cup_k(X \setminus E_{j,k}) \in G_{o^*+}$, and $\cap_j(\cup_k(X \setminus E_{j,k}))$ belong to $G_{o^*++} \subset \mathfrak{G}$, which contradicts the fact that $(X \setminus E) \notin \mathfrak{G}$. Thus there is no $E \in \mathfrak{G}$, such that $(X \setminus E) \notin \mathfrak{G}$.

It remains to verify the condition about countable union. Let $\{E_j\} \subset \bigcup G_o$. We can find o_j, such that $E_j \in G_{o_j}$. As $o = \cup o_j$ is an ordinal, we see that $\{E_j\} \subset G_o$. But then both union and intersection of $\{E_j\}$ should belong to $(G_o)_{\delta\sigma} = G_{o+} \subset \mathfrak{G}$. $\qquad\square$

break

Theorem 4.5. $G_{o_1} = G_{o_2}$ *for any two ordinals greater than* $\aleph_1$, *and so Borel sets coincide with* $G_{\aleph_1}$.

Proof. It is enough to show that $G_{\aleph_1+} = G_{\aleph_1}$, and then use transfinite induction. As before, consider $\{E_j\} \subset G_{\aleph_1}$. Again, as $\aleph_1$ is open, we can find o_j such that $E_j \in G_{o_j}$, and $o_j \prec \aleph_1$ (and thus no more than countable). Thus $o = \cup_j o_j$ is countable, and $\{E_j\} \subset G_o$. The ordinal $o+$ is countable as it differs from o by just one element, and both union and intersection of $\{E_j\}$ belong to $(G_o)_{\delta\sigma} = G_{o+} \subset \bigcup G_{\aleph_1}$. $\qquad\square$

Finally, let us "count" the Borel sets.

Proposition 4.6. $\mathrm{Card}(\mathfrak{B}) = \mathrm{Card}(2^{\mathbb{N}})$ $(= \mathrm{Card}(\mathbb{R}))$.

Remark 4.7. This implies that not all subsets of the real line are Borel.

In order to prove the Proposition we will need the following two lemmas.

Lemma 4.8. $\mathrm{Card}(\mathbb{R} \times \mathbb{R}) = \mathrm{Card}(\mathbb{R})$.

Proof. When it comes to the cardinality, we can think of $\mathbb{R}$ as $2^{\mathbb{N}}$, i.e. the set of binary sequences. This means that to prove the lemma we have to construct a bijection between the set of pairs of binary sequences $\big((a_1, a_2, \ldots), (b_1, b_2, \ldots)\big)$, and the set of binary sequences $(c_1, c_2, \ldots)$. Let $\phi\big((a_1, a_2, \ldots), (b_1, b_2, \ldots)\big) = (a_1, b_1, a_2, b_2, \ldots)$. This is the required bijection. $\qquad\square$

Exercise 4.6. Verify that ϕ is a bijection.

Lemma 4.9. $\mathrm{Card}(\mathbb{R}^{\mathbb{N}}) = \mathrm{Card}(\mathbb{R})$, *where $\mathbb{R}^{\mathbb{N}}$ is a set of all sequences of real numbers.*

Proof. As in the previous case, by replacing $\mathbb{R}$ with $2^{\mathbb{N}}$, we just need to produce a bijection between the set of sequences of binary sequences $((a_1^1, a_2^1, \ldots), (a_1^2, a_2^2, \ldots), \ldots)$ and the set of binary sequences $(b_1, b_2, \ldots)$. Consider a bijection ψ from $\mathbb{N} \times \mathbb{N}$ to $\mathbb{N}$. The map ϕ which sets $b_{\psi(j,k)} = a_k^j$ does the work. $\qquad\square$

Exercise 4.7. Verify that ϕ is a bijection.

Proof of Proposition. Let us observe that the set of all open intervals I has the cardinal number $\mathrm{Card}(\mathbb{R})$. In the next lecture we will prove that $G \subset I_\sigma$. Now observe that if $\mathrm{Card}(H) = \mathrm{Card}(\mathbb{R})$, then $\mathrm{Card}(H_\sigma) = \mathrm{Card}(H_\delta) = \mathrm{Card}(\mathbb{R})$, as the set of countable unions is the natural image of the set of countable sequences of sets from H.

Thus, we can use transfinite induction to show that if $o \preceq \aleph_1$ then G_o has the cardinal number $\mathrm{Card}(\mathbb{R})$.

Indeed, consider the minimal $o \preceq \aleph_1$ for which $\mathrm{Card}(G_o) \neq \mathrm{Card}(\mathbb{R})$. If o is open then G_o is just the union of all preceding sets, and the union of no more than $\aleph_1$ of sets of cardinality $\mathbb{R}$ has the cardinal number $\mathrm{Card}(\mathbb{R})$ (as $\mathrm{Card}(\mathbb{R} \times \mathbb{R}) = \mathrm{Card}(\mathbb{R})$). Now, if o has a maximal element, we know that the set corresponding to the previous ordinal has the cardinal number $\mathrm{Card}(\mathbb{R})$. So, $\mathrm{Card}((G_{o-})_\delta) = \mathrm{Card}(\mathbb{R})$ and $\mathrm{Card}(G_o) = \mathrm{Card}((G_{o-})_{\delta\sigma}) = \mathrm{Card}(\mathbb{R})$. $\qquad\square$

Lecture 5
Measure by a squeeze

We start by recalling the following.

Definition 5.1. A set G is called *open* if for each $x \in G$, there exists $\epsilon > 0$ such that $|y - x| < \epsilon$ implies $y \in G$. A set is called *closed* if it is the complement of an open set.

Exercise 5.1. Show that any union of open sets is open, and that the intersection of any family of closed sets is closed.

Exercise 5.2. Show that intersection of two (and thus finitely many) open sets is open, and that union of two closed sets is closed.

In this lecture we are going to define the Lebesgue measure first on an interval, and then on the real line, so all the sets are subsets of one or the other (in the case it doesn't matter whether we use the line or the interval we denote them as X).

Lemma 5.2. *Each open set $G \subset X$ is a union of no more than countably many disjoint open intervals.*

Proof. Let us first fix a point $x \in G$. If we take $a_x = \sup((X \setminus G) \cap \{y : y < x\})$ and $b_x = \inf((X \setminus G) \cap \{y : y > x\})$ then $a_x, b_x \notin G$, as G is open, and $(a_x, b_x) \subset G$, by the definition of a_x and b_x. This means that every point of G belongs to an open subinterval of G, whose ends don't belong to G. We associate to the point x the interval (a_x, b_x), which contains it. Thus, $G = \bigcup_{x \in G} (a_x, b_x)$. It is easy to see that for two different points $x, y \in G$ the intervals $(a_x, b_x), (a_y, b_y)$ either are disjoint, or coincide.

Exercise 5.3. Prove the last statement.

It remains to show that there are no more than countably many distinct intervals in the union. To do so observe that $G = \bigcup_{x \in G} (a_x, b_x) = \bigcup_{x \in G \cap \mathbb{Q}} (a_x, b_x)$, as each open interval contains a rational point, and $G \cap \mathbb{Q}$ is at most countable.

Exercise 5.4. We have shown that G is a union of no more than countably many open intervals of which each pair either coincide or is disjoint. What remains to prove? Complete the proof of the lemma.

□

Now, we can define (presumably a measure) μ on an open set $G = \bigcup_j (a_j, b_j)$, where the union is disjoint, by letting $\mu(G) = \sum_j (b_j - a_j)$, where the union is taken over the collection of disjoint intervals provided by the lemma above.

Exercise 5.5. In order for μ to be defined it should not depend on which collection of disjoint intervals gives G as the union. Show that the collection of intervals constructed in the lemma 5.2 is a unique one (with the union G).

The defined "measure"[1] μ is monotone in the following sense.

Lemma 5.3. *If $G_1 \subset G_2$ then $\mu(G_1) \leq \mu(G_2)$.*

Proof. Let $G_2 = \cup_j I_j$, where I_j are the disjoint open intervals. Then $G_1 = \cup_j (G_1 \cap I_j)$.

Exercise 5.6. Show that if $G_1 = \cup_k \widetilde{I}_k$ then for every k there exists unique $j(k)$ such that $\widetilde{I}_k \subset G_1 \cap I_{j(k)}$.

It is enough to prove the lemma assuming G_2 to be an interval (a, b). In this case, let $G_1 = \cup_k \widetilde{I}_k$. For each finite disjoint collection $\widetilde{I}_1, \ldots, \widetilde{I}_m \subset (a, b)$ (which can be assumed to be ordered by its position along the line), we see that $\mu(\widetilde{I}_1) + \ldots + \mu(\widetilde{I}_m) = (b_1 - a_1) + \ldots + (b_m - a_m) \leq b_m - a_1 \leq b - a = \mu((a, b))$. Thus, taking the limit, we obtain $\sum_k \mu(\widetilde{I}_k) \leq \mu((a, b))$. □

We now define $\mu(K) = 1 - \mu((0, 1) \setminus K)$ for each closed $K \subset (0, 1)$. While monotonicity of μ on the closed sets follows immediately from the monotonicity of μ on the open sets, it is not clear at all that this monotonicity still holds on the collection of open *and* closed sets.

---

[1] We haven't yet defined μ on σ-algebra, nor have we proven that it is (countably) additive.

Lemma 5.4. *If $G \subset K$ where G is open and K is closed then $\mu(G) \leq \mu(K)$.*

Proof. Let $\widetilde{G} = (0,1) \setminus K$. Then G and $\widetilde{G}$ are disjoint. Consider the representation $G = \cup_j I_j$ and $\widetilde{G} = \cup_k \widetilde{I_k}$, given by the Lemma 5.2. The collection $\{I_j\}_j \cup \{\widetilde{I_k}\}_k$ is a collection of (distinct) disjoint intervals, and so

$$\mu(G) + \mu(\widetilde{G}) = \sum_j |b_j - a_j| + \sum_k |\widetilde{b_k} - \widetilde{a_k}| = \mu(G \cup \widetilde{G}) \leq \mu((0,1)) = 1.$$

Thus, $\mu(K) = 1 - \mu(\widetilde{G}) \geq \mu(G)$. $\qquad\qquad\qquad\qquad\square$

Lemma 5.5. *If $K \subset G$ where G is open and K is closed then $\mu(K) \leq \mu(G)$.*

Proof. Let $G = \bigcup_j I_j$. As K is compact (closed and bounded), and is covered by $\{I_j\}$, one can choose finite subcovering: $K \subset \bigcup_{k=1}^{m} I_{j_k}$, where $I_j = (a_j, b_j)$, and we can assume that the intervals I_{j_k} are numbered in the increasing order, and each contains a point of K. Now let $\widetilde{G} = (0,1) \setminus K = \cup_n \widetilde{I_n}$.

Assume that $a_{j_1} > 0$ and $b_{j_m} < 1$. As $a_{j_1} \notin G$ and so is not a point of K it belongs to its complement and, thus, belong to some of intervals $\{\widetilde{I_n}\}$. Then consider the interval $\widetilde{I_{n_1}}$ which contains a_{j_1}. As there are no points of K in $[0, a_{j_1}]$, the left end $\widetilde{a_{n_1}}$ of the interval $\widetilde{I_{n_1}}$ is 0. On the other hand, as not all points of (a_{j_1}, b_{j_1}) belong to $\widetilde{G}$, we know that $\widetilde{b_{n_1}} < b_{j_1}$. Consider, now the interval $\widetilde{I_{n_2}}$, which contains b_{j_1}. As $[b_{j_1}, a_{j_2}] \subset \widetilde{G}$, the interval $\widetilde{I_{n_2}}$ also contains a_{j_2}. Yet, as not all points of (a_{j_2}, b_{j_2}) are in $\widetilde{G}$, $\widetilde{b_{n_2}} < b_{j_2}$. We continue the construction in this manner, and obtain the finite sequence $\{\widetilde{I_{n_k}}\}_{k=1}^{m+1}$ such that $\widetilde{a_{n_1}} = 0$, $\widetilde{b_{n_{m+1}}} = 1$ and $\widetilde{a_{n_k}} < a_{j_k} < \widetilde{b_{n_k}} < \widetilde{a_{n_{k+1}}} < b_{j_k}$.

Thus, $\mu(\bigcup_{k=1}^{m} I_{j_k}) + \mu(\bigcup_{k=1}^{m+1} \widetilde{I_{n_k}}) > 1$. As $\mu(G) \geq \mu(\bigcup_{k=1}^{m} I_{j_k})$ and $\mu(\widetilde{G}) \geq \mu(\bigcup_{k=1}^{m+1} \widetilde{I_{n_k}})$, this implies $\mu(G) + \mu(\widetilde{G}) > 1$. Thus $\mu(K) = 1 - \mu(\widetilde{G}) < \mu(G)$.

Exercise 5.7. Deal with the cases when $a_{j_1} = 0$ or $b_{j_m} = 1$.

$\square$

Definition 5.6. We introduce the *outer measure* $\mu^*(E) = \inf\{\mu(G) : E \subset G\}$, where G runs through all open supersets of E, and the *inner measure* $\mu_*(E) = \sup\{\mu(K) : K \subset E\}$, where K runs through all closed subsets of E.

The following two lemmas imply that the measures behave reasonably on the open and closed sets.

Lemma 5.7. $\mu^*(G) = \mu_*(G) = \mu(G)$ *for each open set* $G \subset (0,1)$.

Exercise 5.8. Prove the lemma. (Hint: $\mu^*(G) = \mu(G)$ and $\mu_*(G) \leq \mu(G)$ follow from monotonicity. To show that $\mu_*(G) = \mu(G)$ it is enough to construct $K_\epsilon \subset G$, such that $\mu(K_\epsilon)$ approximates $\mu(G)$ to within ϵ. Do it first for the case when G is an interval.)

Corollary 5.8. $\mu^*(K) = \mu_*(K) = \mu(K)$ *for each closed set* $K \subset (0,1)$.

Exercise 5.9. Derive the corollary.

break

While there is no guarantee that either inner or outer measure is actually a measure, there is some hope, as the following is true:

Lemma 5.9. *For an at most countable collection of pairwise disjoint open sets* $\{G_j\}$ *we have* $\mu(\cup_j G_j) = \sum_j \mu(G_j)$.

Exercise 5.10. Show this.

Lemma 5.10. *For a finite collection of pairwise disjoint closed sets $\{K_j\}$ we have $\mu(\cup_j K_j) = \sum_j \mu(K_j)$.*

Proof. The union of finitely many closed sets is closed. So, $\mu(\cup_j K_j) = \mu^*(\cup_j K_j)$. We may choose an open set G such that $\cup_j K_j \subset G$ and $\mu(G) \leq \mu(\cup_j K_j) + \varepsilon$. Now let $2\delta = \min(\text{dist}(K_j, K_m) : m \neq j)$. Let $(K)_\delta = \{X \in (0,1) : \text{dist}(x, \cup_j K_j) < \delta\}$. The set $(K)_\delta$ is open. Let $G \cap (K)_\delta = \cup I_n$ be a union of disjoint intervals. If I_n intersects K_j, then it doesn't intersect any other K_m. Consider $G_j = \bigcup_{I_n \cap K_j \neq \emptyset} I_n$. As the sets G_j consist of different intervals they are disjoint. At the same time $K_j \subset G_j$.

By the Lemma 5.9 and the monotonicity of μ on open/closed sets $\sum \mu(K_j) \leq \sum \mu(G_j) = \mu(\cup G_j) = \mu(G \cap (K)_\delta) \leq \mu(G) \leq \mu(\cup K_j) + \varepsilon$. Thus, the estimate $\sum \mu(K_j) \leq \mu(\cup K_j) + \varepsilon$ holds for all $\varepsilon > 0$ and, so, $\sum \mu(K_j) \leq \mu(\cup K_j)$.

Exercise 5.11. Complete the proof by showing that $\mu(\cup K_j) \leq \sum \mu(K_j)$. (Hint: Pick a collection of open sets such that $K_j \subset G_j$ and $\mu(G_j) \leq \mu(K_j) + \varepsilon$.)

□

Exercise* 5.12. Show that if K_j and K_m are two disjoint (bounded) closed sets, than the distance between them is positive.

Exercise* 5.13. Show that in the previous exercise the boundedness of the sets is essential.

For μ^* and μ_* acting on any set we can prove the following.

Lemma 5.11. *For each finite or countable $\{E_j\}$ we have $\mu^*(\cup_j E_j) \leq \sum_j \mu^*(E_j)$.*

Proof. Let G_j be an open set containing E_j such that $\mu(G_j) \leq \mu^* E_j + \dfrac{1}{2^j}\varepsilon$ Then $G = \cup G_j$ is an open set which contains $\cup E_j$. If we show that $\mu(\cup G_j) \leq \sum \mu(G_j)$, then $\mu^*(\cup E_j) \leq \mu(G) \leq \sum \mu(G_j) \leq \sum \mu^*(E_j) + \varepsilon$. As this can be done for each $\varepsilon > 0$, the desired inequality follows.

33

Let $G = \cup I_k$. Then $\mu(G) = \sum \mu(I_k)$ and for each G_j, $\mu(G_j) = \sum_k \mu(G_j \cap I_k)$ by Lemma 5.9. This means that it is enough to show that
$$\mu(I_k) \leq \sum_j \mu(G_j \cap I_k).$$

Fix an interval $I_k = (a_k, b_k)$. As $G = \cup G_j$, we know that $I_k = \cup(G_j \cap I_k)$. Each $G_j \cap I_k = \bigcup_m I_{j,m}$, so $I_k = \bigcup_{j,m} I_{j,m}$. Consider the interval $[a_k + \delta, b_k - \delta]$ for $0 < \delta < \dfrac{b_k - a_k}{2}$. As it is a compact set covered by open intervals $\{I_{j,m}\}$ we can pick a finite subcovering $\{I_{j(n),m(n)}\}$ such that $[a_k + \delta, b_k - \delta] \subset \bigcup_{n=1}^{N} \{I_{j(n),m(n)}\}$. As in the proof of Lemma 5.5 we can show that $(b_k - \delta) - (a_k + \delta) = (b_k - a_k) - 2\delta \leq \bigcup_{n=1}^{N} \mu(I_{j(n),m(n)}) \leq \sum_{j,m} \mu(I_{j,m}) = \sum_j \mu(G_j \cap I_k)$, i.e. $\mu(I_k) - 2\delta \leq \sum_j \mu(G_j \cap I_k)$. As δ can be an arbitrarily small positive number, this implies $\mu(I_k) \leq \sum_j \mu(G_j \cap I_k)$. $\square$

Lemma 5.12. *For each pairwise disjoint family $\{E_j\}$ we have $\mu_*(\cup_j E_j) \geq \sum_j \mu_*(E_j)$.*

Proof. Let $K_j \subset E_j$ be closed (disjoint) sets such that $\mu(K_j) \geq \mu_*(E_j) - \varepsilon \dfrac{1}{2^j}$. Then for each n the union $\bigcup_{j=1}^{n} K_j$ is a closed subset of $\cup_j E_j$. Thus, $\mu_*(\cup_j E_j) \geq \mu(\bigcup_{j=1}^{n} K_j) = \sum_{j=1}^{n} \mu(K_j) \geq \sum_{j=1}^{n} \mu_*(E_j) - \varepsilon$. If the collection $\{E_j\}$ is infinite, letting $n \to \infty$ gives $\mu_*(\cup_j E_j) \geq \sum_j \mu_*(E_j) - \varepsilon$ (if the collection is finite, take n to be the number of the sets in the collection), for each $\varepsilon > 0$. So, for $\{E_j\}$ we have $\mu_*(\cup_j E_j) \geq \sum_j \mu_*(E_j)$. $\square$

Let us recall the following:

Definition 5.13. A collection $\mathfrak{A}$ of subsets of X is called a *σ-algebra* if

1. $X \in \mathfrak{A}$;

2. If $E \in \mathfrak{A}$, then $(X \setminus E) \in \mathfrak{A}$;

34

3. For each collection $\{E_j\}_{j=1}^{\infty} \subset \mathfrak{A}$, we have $(\cup E_j) \in \mathfrak{A}$.

In order to define a measure on the σ-algebra generated by the open sets, we may define a measure on some σ-algebra, which contains the open sets (and then take the restriction to the smaller σ-algebra).

Proposition 5.14. *The collection of subsets of* $(0,1)$ *for which* $\mu^*(E) = \mu_*(E)$ *is a* σ*-algebra.*

Thus, considering μ^* (or μ_*) on the collection of the sets on which inner and outer measures coincide we introduce a measure on a σ-algebra which contains the open sets, since $\mu^* = \mu_* = \mu$ for the open sets. The measure defined in this way is called Lebesgue measure. Notice, that the defined σ-algebra contains Borel sets, but it also contains other sets. All sets of the σ-algebra on which Lebesgue measure is defined are called *Lebesgue measurable*.

For the prove of the proposition it is more useful to use an equivalent definition of σ-algebra.

Definition 5.15. A collection $\mathfrak{A}$ of subsets of X is called a σ-*algebra* if

1. $X \in \mathfrak{A}$;

2. If $E_1, E_2 \in \mathfrak{A}$, then $(E_1 \setminus E_2) \in \mathfrak{A}$;

3. For each disjoint collection $\{E_j\}_{j=1}^{\infty} \subset \mathfrak{A}$, we have $(\cup E_j) \in \mathfrak{A}$.

Exercise 5.14. Show that the two definitions are equivalent.

Proof of the Proposition. The property (1) is trivial. The property (3) is a combination of Lemma 5.11 and Lemma 5.12.

To show the property (2) observe that $\mu^*(E) = \mu_*(E)$ is equivalent to the fact that for each $\varepsilon > 0$ there exists $K \subset E \subset G$ (where K is closed and G is open), such that $\mu(G \setminus K) < \varepsilon$ (observe that $G \setminus K$ is open, so μ is well-defined).

Now, pick $K_1 \subset E_1 \subset G_1$ and $K_2 \subset E_2 \subset G_2$ such that $\mu(G_1 \setminus K_1) + \mu(G_2 \setminus K_2) \leq \varepsilon$. Then $K_1 \setminus G_2 \subset E_1 \setminus E_2 \subset G_1 \setminus K_2$. And $\mu((G_1 \setminus K_2) \setminus (K_1 \setminus G_2)) \leq \mu((G_1 \setminus K_1) \cup (G_2 \setminus K_2))$. The latter, by

35

Lemma 5.11, is not greater than $\mu(G_1 \setminus K_1) + \mu(G_2 \setminus K_2) \leq \varepsilon$, which completes the proof. $\qquad\square$

Exercise 5.15. Show that $(G_1 \setminus K_2) \setminus (K_1 \setminus G_2) \subset (G_1 \setminus K_1) \cup (G_2 \setminus K_2)$.

By the combination of Lemma 5.11 and Lemma 5.12, the measure μ defined on the sets for which $\mu^*(E) = \mu_*(E)$ is countably additive .

Exercise 5.16. Show that the measure is invariant with respect to a shift inside the interval.

Exercise 5.17. Extend the definition of the measure to $\mathbb{R}$.

Exercise 5.18. Show that not all Lebesgue measurable sets are Borel. (Hint: We have shown that $\mathrm{Card}(\mathfrak{B}) = \mathrm{Card}(\mathbb{R})$. Show that cardinality of Lebesgue measurable sets is at least $\mathrm{Card}(2^{\mathbb{R}})$. To do so find a set of $\mathrm{Card}(\mathbb{R})$ which has Lebesgue measure zero and consider all its subsets.)

Lecture 6
A squeeze from one side

Let us consider the $\frac{1}{3}$-Cantor set $\mathcal{C}$, i.e. the set obtained by removing the (open) middle third from the interval $[0, 1]$, and then middle thirds from each of the two intervals which are left, and so on. Observe that $\mathcal{C}$ is closed (because its complement is open).

Exercise* 6.1. The $\frac{1}{3}$-Cantor set can be described as the set of points, which can be represented in base 3 presented by an (infinite) expansion which includes only "0" and "2". Show that the two definitions are equivalent.

We can find the Lebesgue measure of $\mathcal{C}$ by the definition of μ for a closed set. Its complement is an open set, which consists of one interval of length $\frac{1}{3}$, two intervals of length $\frac{1}{9}$, four intervals of the length $\frac{1}{3^3}$, etc. Thus $\mu(\mathcal{C}) = 1 - \sum_{j=0} \frac{2^j}{3^{j+1}} = 1 - \frac{1}{3}\frac{1}{1-2/3} = 0$. Another way to obtain the same result is to observe that $\mathcal{C}$ is a subset of two intervals of length $\frac{1}{3}$, and thus has the measure not greater than $\frac{2}{3}$. It is also a subset of four intervals of length $\frac{1}{9}$, and thus has measure not greater than $\frac{4}{9} = (\frac{2}{3})^2$. If we continue the estimates in this way we see that $\mu(\mathcal{C}) \leq (\frac{2}{3})^n \to 0$. Actually this proves that $\mu^*(\mathcal{C}) = 0$, but as $\mathcal{C}$ is closed $\mu^*(\mathcal{C}) = \mu(\mathcal{C})$. This is the route to define a measure we take today: First define an outer measure $\mathcal{H}^*_\alpha$, and then set $\mathcal{H}_\alpha = \mathcal{H}^*_\alpha$ on the sets which we deem measurable.

Let us for a number α and $\delta > 0$ define $\mathcal{H}^\delta_\alpha(E) = \inf\{\sum_j |I_j|^\alpha : \cup I_j \supset E, |I_j| < \delta\}$, where $\{I_j\}$ is a countable collection of open intervals which covers E, and $|I_j|$ is the length of the interval I_j. When δ decreases, we take the infimum over smaller collection of numbers, and thus $\mathcal{H}^\delta_\alpha(E)$ is increasing as a function of δ. Thus, the limit $\mathcal{H}^*_\alpha(E) = \lim_{\delta \to 0} \mathcal{H}^\delta_\alpha(E)$ exists, but is not necessarily finite.

Exercise 6.2. Show that if $\alpha = 1$ then for each $\delta > 0$ we have $\mathcal{H}^\delta_\alpha(E) =$

$\mu^*(E)$, where the μ^* is the outer measure we have already defined in Lecture 5. Thus, $\mathcal{H}_1^*(E) = \mu^*(E)$.

Exercise 6.3. Show that if $\alpha > 1$ then for each $\delta > 0$ we have $\mathcal{H}_\alpha^\delta(E) = 0$. Thus, $\mathcal{H}_\alpha^*(E) = 0$ for each $E \subset \mathbb{R}$.

Exercise 6.4. Show that if $\alpha < 1$ then $\mathcal{H}_\alpha^*((0,1)) = +\infty$. (Hint: The function $f(t) = t^\alpha, t > 0$ is subadditive, i.e. $f(s+t) \leq f(s) + f(t)$.) Similarly, $\mathcal{H}_\alpha^*(G) = +\infty$ for each open set G.

Exercise 6.5. Show that $\mathcal{H}_\alpha^*$ is monotone, i.e. if $E_1 \subset E_2$ then $\mathcal{H}_\alpha^*(E_1) \leq \mathcal{H}_\alpha^*(E_2)$.

Lemma 6.1. *For any sets E_1, E_2 we have*

$$\mathcal{H}_\alpha^*(E_1 \cup E_2) \leq \mathcal{H}_\alpha^*(E_1) + \mathcal{H}_\alpha^*(E_2).$$

Proof. Let for some $\varepsilon > 0$ and a fixed δ the collection $\{I_{j,1}\}$, $|I_{j,1}| \leq \delta$, be a covering of E_1 such that $\sum_j |I_{j,1}|^\alpha \leq \mathcal{H}_\alpha^\delta(E_1) + \varepsilon$ and the collection $\{I_{j,2}\}$, $|I_{j,2}| \leq \delta$ be a covering of E_2 such that $\sum_j |I_{j,2}|^\alpha \leq \mathcal{H}_\alpha^\delta(E_2) + \varepsilon$. Then the union $\{I_{j,1}\} \cup \{I_{j,2}\}$ is a covering of $E_1 \cup E_2$ and $\mathcal{H}_\alpha^\delta(E_1 \cup E_2) \leq \sum_j |I_{j,1}|^\alpha + \sum_j |I_{j,2}|^\alpha \leq \mathcal{H}_\alpha^\delta(E_1) + \mathcal{H}_\alpha^\delta(E_2) + 2\varepsilon$. So, $\mathcal{H}_\alpha^\delta(E_1 \cup E_2) \leq \mathcal{H}_\alpha^\delta(E_1) + \mathcal{H}_\alpha^\delta(E_2) + 2\varepsilon$ for each $\varepsilon > 0$ and so

$$\mathcal{H}_\alpha^\delta(E_1 \cup E_2) \leq \mathcal{H}_\alpha^\delta(E_1) + \mathcal{H}_\alpha^\delta(E_2) \leq \mathcal{H}_\alpha^*(E_1) + \mathcal{H}_\alpha^*(E_2).$$

It remains to let $\delta \to 0$. $\qquad\square$

Exercise 6.6. Show the same inequality for a countable union.

Corollary 6.2. *For each $E \subset \mathbb{R}$ and $A \subset \mathbb{R}$ we have*

$$\mathcal{H}_\alpha^*(E) \leq \mathcal{H}_\alpha^*(E \cap A) + \mathcal{H}_\alpha^*(E \setminus A).$$

Definition 6.3. We say that a set A is *Hausdorff-α measurable* if for each E we have $\mathcal{H}_\alpha^*(E) = \mathcal{H}_\alpha^*(E \cap A) + \mathcal{H}_\alpha^*(E \setminus A)$.

Remark 6.4. Notice that we do not say that a measurable set has a finite measure.

Proposition 6.5. *The Hausdorff-α measurable sets form a σ-algebra.*

Proof. We give the proof for the equivalent definition of σ-algebra, as we did the previous time.

Definition 6.6. A collection $\mathfrak{A}$ of subsets of X is called σ-*algebra* if

1. $X \in \mathfrak{A}$;

2. If $A_1, A_2 \in \mathfrak{A}$, then $(A_1 \setminus A_2) \in \mathfrak{A}$;

3. For each collection of disjoint sets $\{A_j\}_{j=1}^{\infty} \subset \mathfrak{A}$, we have $(\cup A_j) \in \mathfrak{A}$.

Now, the property 1 is trivial. Property 2 follows from the observation that if $A_1, A_2 \in \mathfrak{A}$ then for each E we have

$$\mathcal{H}_\alpha^*(E) = \mathcal{H}_\alpha^*(E \cap A_1) + \mathcal{H}_\alpha^*(E \setminus A_1),$$

so

$$\mathcal{H}_\alpha^*(E) == \mathcal{H}_\alpha^*((E \cap A_1) \cap A_2) + \mathcal{H}_\alpha^*((E \cap A_1) \setminus A_2) + \mathcal{H}_\alpha^*(E \setminus A_1).$$

By Lemma 6.1, for the first and the third term of the right hand side we have $\mathcal{H}_\alpha^*((E \cap A_1) \cap A_2) + \mathcal{H}_\alpha^*(E \setminus A_1) \geq \mathcal{H}_\alpha^*(E \setminus (A_1 \setminus A_2))$.

Exercise 6.7. Show that $(E \cap A_1 \cap A_2) \cup (E \setminus A_1) = E \setminus (A_1 \setminus A_2)$ and $((E \cap A_1) \setminus A_2) = E \cap (A_1 \setminus A_2)$.

We see, thus, that $\mathcal{H}_\alpha^*(E) \geq \mathcal{H}_\alpha^*(E \cap (A_1 \setminus A_2)) + \mathcal{H}_\alpha^*(E \setminus (A_1 \setminus A_2))$. The opposite inequality follows from Lemma 6.1, which proves the property 2.

As $A_1 \cup A_2 = X \setminus ((X \setminus A_1) \setminus A_2)$, the properties 1 and 2 imply that a union of two (and thus finitely many) measurable sets is measurable.

So, it remains to deal with the case when the collection $\{A_j\}$ is countable. The Lemma 6.1 shows that $\mathcal{H}_\alpha^*(E) < \mathcal{H}_\alpha^*(E \cap (\cup A_j)) + \mathcal{H}_\alpha^*(E \setminus (\cup A_j))$. On the other hand, for each n, as the sets are disjoint and measurable, $\mathcal{H}_\alpha^*(E) = \mathcal{H}_\alpha^*(E \cap A_1) + \mathcal{H}_\alpha^*(E \setminus A_1) = \mathcal{H}_\alpha^*(E \cap A_1) + \mathcal{H}_\alpha^*(E \cap A_2) +$

39

$$\mathcal{H}_\alpha^*(E \setminus (A_1 \cup A_2)) = \ldots = \sum_{j=1}^{n} \mathcal{H}_\alpha^*(E \cap A_j) + \mathcal{H}_\alpha^*(E \setminus (\bigcup_{j=1}^{n} A_j)).$$ By monotonicity of $\mathcal{H}_\alpha^*$ we see that $\mathcal{H}_\alpha^*(E) \geq \sum_{j=1}^{n} \mathcal{H}_\alpha^*(E \cap A_j) + \mathcal{H}_\alpha^*(E \setminus (\bigcup_{j=1}^{\infty} A_j)).$ Now, we can pass to the limit in n, so that $\mathcal{H}_\alpha^*(E) \geq \sum_{j=1}^{\infty} \mathcal{H}_\alpha^*(E \cap A_j) + \mathcal{H}_\alpha^*(E \setminus (\bigcup_{j=1}^{\infty} A_j)).$ Using the improved version of Lemma 6.1 (which you have proved in Exercise 6.6) on the first term in the right-hand side, we see that

$$\mathcal{H}_\alpha^*(E) \geq \mathcal{H}_\alpha^*(E \cap (\cup A_j)) + \mathcal{H}_\alpha^*(E \setminus (\cup A_j)),$$

which completes the proof. □

As before it remains to prove that Borel sets are measurable. As we know that the measurable sets form a σ-algebra, it is sufficient to show that all the open sets are measurable. (Notice that to be measurable a set doesn't need to have a finite measure.) As each open set is a countable union of intervals, it is even enough to show that all open intervals are measurable.

Lemma 6.7. *Each interval $I = (a, b)$ is $\mathcal{H}_\alpha$-measurable.*

Proof. Consider an arbitrary set E. The Lemma 6.1 tells us that $\mathcal{H}_\alpha^*(E) \leq \mathcal{H}_\alpha^*(E \cap I) + \mathcal{H}_\alpha^*(E \setminus I)$.

To see the opposite inequality, we may assume that $\mathcal{H}_\alpha^* < +\infty$. Fix $\delta, \varepsilon > 0$. Take a covering $\{I_j\}$ of E such that $|I_j| \leq \delta$ and $\sum_j |I_j|^\alpha < \mathcal{H}_\alpha^\delta(E) + \varepsilon$. Let $\mathcal{J}$ be the collection of all indices of the intervals $\{I_j\}$ which intersect (a, b). Then $\{I_j\}_{j \in \mathcal{J}}$ is a covering of $E \cap I$. At the same time as the lengths of the intervals do not exceed δ the collection $\{I_j\}_{j \notin \mathcal{J}}$ is a covering of $E \setminus (a - \delta, b + \delta)$. Thus, the collection

$$\{I_j\}_{j \notin \mathcal{J}} \cup \{(a - \delta, a), (a - \delta/2, a + \delta/2), (b, b + \delta), (b - \delta/2, b + \delta/2)\}$$

is a covering of $E \setminus I$. This implies that $\mathcal{H}_\alpha^\delta(E \cap I) + \mathcal{H}_\alpha^\delta(E \setminus I) \leq \sum_j |I_j|^\alpha + 4\delta^\alpha \leq \mathcal{H}_\alpha^\delta(E) + \varepsilon + 4\delta^\alpha$. As ε can be any positive number we see that $\mathcal{H}_\alpha^\delta(E \cap I) + \mathcal{H}_\alpha^\delta(E \setminus I) \leq \mathcal{H}_\alpha^\delta(E) + 4\delta^\alpha$. When we let $\delta \to 0$ the inequality becomes $\mathcal{H}_\alpha^*(E) \geq \mathcal{H}_\alpha^*(E \cap I) + \mathcal{H}_\alpha^*(E \setminus I)$, which completes the proof. $\qquad\square$

We define $\mathcal{H}_\alpha(E) = \mathcal{H}_\alpha^*(E)$ for all Hausdorff-α measurable sets, and so it is defined on the Borel sets. Now, if A_1 and A_2 are two disjoint measurable sets then, as A_1 is measurable $\mathcal{H}_\alpha(A_1 \cup A_2) = \mathcal{H}_\alpha((A_1 \cup A_2) \cap A_1) + \mathcal{H}_\alpha((A_1 \cup A_2) \setminus A_1) = \mathcal{H}_\alpha(A_1) + \mathcal{H}_\alpha(A_2)$, so the measure is finitely additive.

If $\{A_j\}$ is a countable collection of disjoint measurable sets, then $\mathcal{H}_\alpha(\cup A_j) \leq \sum \mathcal{H}_\alpha(A)$ by the improved Lemma 6.1. On the other hand by monotonicity, for each n we have $\sum_{j=1}^{n} \mathcal{H}_\alpha(A_j) = \mathcal{H}_\alpha(\bigcup_{j=1}^{n} A_j) \leq \mathcal{H}_\alpha(\cup A_j)$ and by letting $n \to \infty$ we obtain $\mathcal{H}_\alpha(\cup A_j) \geq \sum \mathcal{H}_\alpha(A)$, which completes the proof of countable additivity of the defined measure $\mathcal{H}_\alpha$.

Break

Most of what we have done before the break is not specific for $\mathbb{R}$ (one-dimensional space).

Exercise* 6.8. Check which results and to what extent can be generalized[1] to $\mathbb{R}^n$. (Hint: The argument that the Borel sets are measurable requires that $n - \alpha < 1$.)

A proof that the Borel sets are Hausdorff-α measurable which holds for each dimension requires a bit more work than the proof for one dimension. It is implied by the following:

Lemma 6.8. *Each closed set K is Hausdorff-α measurable.*

---

[1]It is not clear what should be the analogue of an interval in the multidimensional settings, but the most commonly-done thing is to consider arbitrary sets S, with $|I|$ being replaced by $\text{diam}(S) - \sup\{\text{dist}(x,y) . x, y \in S\}$, though sometimes one uses only the balls as the analogues of the intervals, and the diameter as the analogue of the length.

Proof. Observe that if $\mathrm{dist}(E_1, E_2) > 0$ then $\mathcal{H}^*_\alpha(E_1 \cup E_2) = \mathcal{H}^*_\alpha(E_1) + \mathcal{H}^*_\alpha(E_2)$.

Exercise 6.9. Show this.

Let $A_j = \{x : \frac{1}{2^j} < \mathrm{dist}(x, K)\}$ and $B_j = A_{j+1} \setminus A_j$. Then $\mathrm{dist}(B_j, B_{j+2}) \geq \frac{1}{2^{j+2}} > 0$, and so for each E we have $\mathcal{H}^*_\alpha(E \cap (B_j \cup B_{j+2})) = \mathcal{H}^*_\alpha(E \cap B_j) + \mathcal{H}^*_\alpha(E \cap B_{j+2})$.

Exercise 6.10. Show that for each n we have

$$\mathcal{H}^*_\alpha \left(E \cap \left(\bigcup_{k=0}^n B_{j+2k} \right) \right) = \sum_{k=0}^n \mathcal{H}^*_\alpha(E \cap B_{j+2k}).$$

Assume that $\mathcal{H}^*_\alpha(E) < \infty$. Then for each n we have $\sum_{k=0}^n \mathcal{H}^*_\alpha(E \cap B_{2k}) = \mathcal{H}^*_\alpha \left(E \cap \left(\bigcup_{k=0}^n B_{2k} \right) \right) \leq \mathcal{H}^*_\alpha(E) < \infty$ and so the series $\sum_{k=0}^n \mathcal{H}^*_\alpha(E \cap B_{2k})$ is convergent. Similarly, the series $\sum_{k=0}^n \mathcal{H}^*_\alpha(E \cap B_{2k+1})$ is convergent, and, thus, $\sum_{k=0}^n \mathcal{H}^*_\alpha(E \cap B_k)$ is convergent.

Now, as K is closed, we have $X \setminus K = A_n \cup (\bigcup_{k \geq n} B_k)$ for each n. Thus, by monotonicity and the improved Lemma 6.1, we have

$$\mathcal{H}^*_\alpha(E \cap A_n) \leq \mathcal{H}^*_\alpha(E \setminus K)$$
$$\leq \quad \mathcal{H}^*_\alpha(E \cap A_n) + \mathcal{H}^*_\alpha \left(E \cap \left(\bigcup_{k \geq n} B_k \right) \right)$$
$$\leq \quad \mathcal{H}^*_\alpha(E \cap A_n) + \sum_{k=n}^\infty \mathcal{H}^*_\alpha(E \cap B_k).$$

If $\mathcal{H}^*_\alpha(E) < \infty$, the second term in the last estimate is the tail of a convergent series, and so tends to zero when $n \to \infty$. This means that $\mathcal{H}^*_\alpha(E \cap A_n) \to \mathcal{H}^*_\alpha(E \setminus K)$ when $n \to \infty$.

If $\mathcal{H}_\alpha^*(E) < \infty$, then $\mathcal{H}_\alpha^*(E) \geq \mathcal{H}_\alpha^*((E \cap A_n) \cup (E \cap K)) = \mathcal{H}_\alpha^*(E \cap A_n) + \mathcal{H}_\alpha^*(E \cap K) \to \mathcal{H}_\alpha^*(E \setminus K) + \mathcal{H}_\alpha^*(E \cap K)$. The opposite inequality, $\mathcal{H}_\alpha^*(E) \leq \mathcal{H}_\alpha^*(E \setminus K) + \mathcal{H}_\alpha^*(E \cap K)$ follows from Lemma 6.1.

Exercise 6.11. Explain what happens when $\mathcal{H}_\alpha^*(E) = \infty$.

$\square$

Self-similar sets

Let us again consider the $\frac{1}{3}$-Cantor set $\mathcal{C}$. When we remove the middle $\frac{1}{3}$ the rest of the interval (including $\mathcal{C}$) can be covered by two intervals of length arbitrary close to $\frac{1}{3}$. I.e. $\mathcal{H}_\alpha^{\frac{1}{2}}(\mathcal{C}) \leq 2(\frac{1}{3})^\alpha$. In the same way we see on the next step of the construction that $\mathcal{H}_\alpha^{\frac{1}{4}}(\mathcal{C}) \leq 4(\frac{1}{3^2})^\alpha$. As we continue we obtain $\mathcal{H}_\alpha^{\frac{1}{2^n}}(\mathcal{C}) \leq 2^n(\frac{1}{3^n})^\alpha = (\frac{2}{3^\alpha})^n$. This estimate shows us that if $\alpha > \log_3(2) = \dfrac{\ln(2)}{\ln(3)}$, then $\mathcal{H}_\alpha(\mathcal{C}) = 0$, as well as $\mathcal{H}_{\frac{\ln(2)}{\ln(3)}}(\mathcal{C}) \leq 1$. We also can suspect that our choices of the covering were close to optimal. Yet, it is not easy to show that no other covering would be better.

Exercise* 6.12. Evaluate $\mathcal{H}_{\frac{\ln(2)}{\ln(3)}}(\mathcal{C})$.

We can generalize the Cantor set by using its self-similarity.

Definition 6.9. A map $s : \mathbb{R}^n \mapsto \mathbb{R}^n$ is called a *similitude* if $\text{dist}(s(x), s(y)) = r \cdot \text{dist}(x, y)$ for some constant r (we will call r the *coefficient* of the similitude s).

Exercise* 6.13. Describe all similitudes of $\mathbb{R}^n$.

Definition 6.10. A compact (closed and bounded) set K is called *self-similar* if there exists a finite collections of similitudes $\{s_j\}_{j=1}^n$ such that $K = \cup s_j(K)$.

The definition itself provides us with a tool to estimate the Hausdorff measure of a self-similar K. Consider a set collection which contains one open ball $B_0 = \{B\}$, such that $K \subset B$. Then, consider $B_1 = \{s_1(B), \dots, s_n(B)\}$. It is a covering of K by n open balls such

43

that $\sum_{j=1}^{n} \operatorname{diam}(s_j(B))^\alpha = (\sum_{j=1}^{n} r_j^\alpha)\operatorname{diam}(B)^\alpha$. If $(\sum_{j=1}^{n} r_j^\alpha) < 1$, by repeating this operation we obtain covering of K by smaller balls with summary α-powers of their diameters tending to zero. We have proven the following.

Proposition 6.11. *If K is a self-similar set, $\{s_j\}_{j=1}^{n}$ is a collection of similarities with coefficients $\{r_j\}_{j=1}^{n}$ such that $K = \bigcup_{j=1}^{n} s_j(K)$, and $\alpha > 0$ is such that $\sum_{j=1}^{n} r_j^\alpha < 1$, then $\mathcal{H}_\alpha(K) = 0$.*

Similarly to the Cantor set it is a bit more complicated to say what happens if $\sum_{j=1}^{n} r_j^\alpha \geq 1$. This we will do in Lecture 8.

As the class of self-similar sets is interesting for Geometric Measure Theory, it is interesting to observe that there are many of them.

Proposition 6.12. *For each collection of similarities $\{s_j\}_{j=1}^{n}$, with coefficients less than 1, there exists a unique compact set K such that $K = \bigcup_{j=1}^{n} s_j(K)$.*

Proof. To show this we consider the metric[2] space $\mathcal{M}$ where points are compact subsets of $\mathbb{R}^n$ and $\operatorname{dist}_M(K_1, K_2)$ equals

$$\max\{\sup\{\operatorname{dist}(x, K_2) : x \in K_1\}, \sup\{\operatorname{dist}(K_1, y) : y \in K_2\}\}.$$

Exercise 6.14. Show that the defined distance $\operatorname{dist}(K_1, K_2)$ is a metric.[3]

Then to the collection of similitudes corresponds a map $S : \mathcal{M} \to \mathcal{M}$, such that $S(K) = \cup s_j(K)$. To show that for some K we have $K = \cup s_j(K)$ is the same as to show that $K = S(K)$.

Exercise 6.15. Show that if the coefficients of the similitudes are less than 1 then S is a contraction, i.e. $\operatorname{dist}_M(S(K_1), S(K_2)) \leq c\cdot\operatorname{dist}_M(K_1, K_2)$, where $c < 1$.

---

[2] It is called the Hausdorff metric
[3] That is, show that $\operatorname{dist}(K, M) = 0$ iff $K = M$; that it is symmetric, and that the triangle inequality holds.

Now, by convergence of a geometric series we see that the sequence $S^n(K)$ is a Cauchy sequence for any initial non-empty compact set K. If this sequence has a limit $\widehat{K}$ then (as it is in any theorem of a fixed point for a contraction) $\widehat{K} = S(\widehat{K})$.

Thus, we need only show that in $\mathcal{M}$ each Cauchy sequence has a limit, i.e. that $\mathcal{M}$ is a complete metric space. Let $\{K_n\}$ be a Cauchy sequence, and K be the set of all points p such that there exists a sequence $p_{n_j} \in K_{n_j}$, such that $p = \lim p_{n_j}$.

First, observe that K is bounded.

If $q = \lim q_k$, where $q_k \in K$, and so there exists $\{p_{k,n_j}\}$ such that $p_{k,n_j} \in K_{n_j}$ and $p_{k,n_j} \to q_k$. We can pick m_k, such $\operatorname{dist}(p_{k,m_k}, q_k) <$ $\operatorname{dist}(q_k, q)$ and $m > k$, so that $p_{k,m_k} \to q$ and $p_{k,m_k} \in K_{m_k}$. Thus the set K is closed. As a closed and bounded subset of $\mathbb{R}^n$, the set K is compact and, so, is a point in $\mathcal{M}$. It remains to show that $K_n \to K$ in $\mathcal{M}$.

Fix $\varepsilon > 0$. Consider N such that $\operatorname{dist}_M(K_n, K_m) < \varepsilon$ for all $m, n \geq N$. Let $n > N$. For each point $x \in K_n$ there exists a point $x_m \in K_m$, $m > n$ such that $\operatorname{dist}(x_m, x) < \varepsilon$. The sequence $\{x_m\}$ is bounded and thus has a limit point $\widetilde{x}$ (a point in $\mathbb{R}^n$. By the definition of K, the point $\widetilde{x}$ belongs to K and by the choice of $\{x_m\}$ the distance $\operatorname{dist}_M(\widetilde{x}, x) \leq \varepsilon$, i.e. $\operatorname{dist}(x, K) \leq \varepsilon$ for each $x \in K_n$ as soon as $n > N$. Thus, $\sup_{x \in K_n} \{\operatorname{dist}(x, K)\} \leq \varepsilon$.

On the other hand, if $\widetilde{x} \in K$, then there exists $m > N$ and $x_m \in K_m$, such that $\operatorname{dist}(\widetilde{x}, x_m) < \varepsilon$. At the same time $\operatorname{dist}(x_m, K_n) < \varepsilon$ for all $n > N$ and the triangle inequality gives us $\operatorname{dist}(\widetilde{x}, K_n) < 2\varepsilon$ for each $n > N$, i.e. $\sup_{\widetilde{x} \in K} \{\operatorname{dist}(\widetilde{x}, K_n)\} \leq 2\varepsilon$. Together with the previous estimate this gives us $\operatorname{dist}_M(K_n, K) < 2\varepsilon$, for all $n > N$. As ε is an arbitrary positive number this completes the proof that the Cauchy sequence has a limit, and thus that the metric space of compacts is complete.

$\square$

Lecture 7
Three covering theorems

As you have seen before, it is important to have some way to handle coverings of a set by sets of a small diameter or by small balls. In the field of Geometric Measure Theory there are three theorems which are very useful for this purpose.

Let us first observe that covering a set by balls is not the same as covering the set by sets of an arbitrary shape (an example is a regular triangle - its diameter is less than the radius of the minimal ball which covers it). On the other hand if the collection $\{A_j\}$ is a covering of a set E, then every set A_j of the collection can be covered by a ball of diameter at most twice that of the set A_j. This means that along with $\mathcal{H}_\alpha$ we can consider $\mathcal{S}_\alpha$ - the spherical measure, i.e. the measure obtained in the same manner as the Hausdorff measure with the diameters of the covering sets being replaced by radii of the covering balls.

Exercise 7.1. Show that Borel sets are $\mathcal{S}_\alpha$-measurable.

Exercise 7.2. Show that for every Borel set E we have $\mathcal{H}_\alpha(E) \leq \mathcal{S}_\alpha(E) \leq 2^\alpha \mathcal{H}_\alpha(E)$.

5r-covering theorem

We denote the radius of a ball B by $r(B)$, and we denote the ball with the same center and radius $t \times r(B)$ by tB.

Theorem 7.1. *Let $\mathcal{B}$ be a collection of closed balls, such that $sup\{r(B) : B \in \mathcal{B}\} < \infty$. Then there exists an at most countable subcollection $\{B_j\} \subset \mathcal{B}$ of disjoint balls such that $\bigcup_{B \in \mathcal{B}} B \subset \bigcup_j 5B_j$.*

Proof. Let $R = sup\{r(B) : B \in \mathcal{B}\}$. Consider a maximal under inclusion pairwise disjoint collection $\mathcal{J}_1$ of balls from $\mathcal{B}$ of radius at least $\frac{1}{2}R$ (— this is possible, because the family of all collections of disjoint balls from $\mathcal{B}$ is ordered by inclusion in a way which allow us to use Zorn's lemma).

47

This collection is obviously countable. Remove from $\mathcal{B}$ all the balls which intersect any of the balls of $\mathcal{J}_1$ and call the set of remaining balls $\mathcal{B}_1$. It is clear that $sup\{r(B) : B \in \mathcal{B}_1\} \le \frac{1}{2}R$, for if there were any ball of radius $R/2$ or greater we could add it to the collection $\mathcal{J}_1$, thus increasing it. We continue the construction by choosing $\mathcal{J}_2$ as a maximal collection of disjoint balls of $\mathcal{B}_1$, and $\mathcal{B}_2$ as the collection of the balls of $\mathcal{B}_1$ which do not intersect any of $\mathcal{J}_2$. Repeat the process infinitely many times (or, more precisely, use Hausdorff maximal principle on the partially ordered set of possible constructions). The resulting set $\{B_j\} = \cup \mathcal{J}_k$ is obviously countable.

Consider each ball $B \in \mathcal{B}$. Let k be such that $2^{-k}R < r(B) \le 2^{-k+1}R$. Then after the step k the ball B is not in the sub-collection $\mathcal{B}_k$ (as $sup\{r(B) : B \in \mathcal{B}_k\} \le \frac{1}{2^k}R$), i.e. it was removed at some step m, $m \le k$. A ball is removed from the collection in two ways - either it is a ball of $\mathcal{J}_m$ or it intersects one of the balls in $\mathcal{J}_m$. In the first case, $B \in \{B_j\}$, in the second case there is a ball $B^* \in \mathcal{J}_m$, such that $B \cap B^* \ne \emptyset$. By the choice of $\mathcal{J}_m$, $2^{-m}R \le r(B^*)$, so $r(B) \le 2r(B^*)$, and it is easy to see that, as the two balls intersect each other, we have $B \subset 5B^*$.

Now, we see that $\bigcup_{B \in \mathcal{B}} B \subset \bigcup_j 5B_j$ as each ball in the first union is inside of some ball from the second union. $\qquad \square$

Besicovitch's covering theorem

Theorem 7.2. [1] *For every dimension n there exist two integers $P(n)$ and $Q(n)$, such that if A is a subset of $\mathbb{R}^n$ and $\mathcal{B}$ is a collection of closed balls, such that for every $x \in A$ there exists $B \in \mathcal{B}$ which has its center at x and*[2] $sup\{r(B) : B \in \mathcal{B}\} < \infty$, *then the following is true.*

1. *It is possible to choose an at most countable collection $\widetilde{\mathcal{B}} \subset \mathcal{B}$, such that $A \subset \bigcup_{B \in \widetilde{\mathcal{B}}} B$ and each point of $\mathbb{R}^n$ is covered by no more than $P(n)$ balls.*

---

[1] Often the first part of the theorem alone is called Besicovitch's covering theorem
[2] Usually one assumes that the set A is bounded, but we do it more generally.

2. *It is possible to divide the collection $\widetilde{\mathcal{B}}$ into $Q(n)$ parts $\widetilde{\mathcal{B}} = \bigcup\limits_{j=1}^{Q(n)} \widetilde{\mathcal{B}}_j$ in such a way that each $\widetilde{\mathcal{B}}_j$ consists of disjoint balls.*

Proof. First, denote $M := \sup\{r(B) : B \in \mathcal{B}\}$.

Take $\mathcal{C}_1$ to be a maximal (by inclusion) sub-collection of balls from $\mathcal{B}$ of radius at least $M/2$, such that no ball in $\frac{1}{2}\mathcal{C}_1 = \{\frac{1}{2}B : B \in \mathcal{C}_1\}$ contains the center of another ball in $\frac{1}{2}\mathcal{C}_1$.

Exercise 7.3. Carefully apply Zorn's lemma to do so.

Observe that if two balls do not contain each other's centers then the balls with the same centers but radii twice as small are disjoint. This means that $\frac{1}{4}\mathcal{C}_1$ is a collection of disjoint balls. In particular that means that $\mathcal{C}_1$ is no more than countable.

Let $x \in A_1 = A \setminus (\bigcup\limits_{B \in \mathcal{C}_1} B)$ be the center of a ball B of radius greater than $M/2$ (the radius still has to be less than M). Then, as $x \notin \bigcup\limits_{B \in \mathcal{C}_1} B$, the distance from x to the center of each ball of $\mathcal{C}_1$ is at least $M/2$ and, thus, adding B to $\mathcal{C}_1$ wouldn't violate the condition with respect to which $\mathcal{C}_1$ is maximal. This contradiction shows that if $\mathcal{B}_1 \subset \mathcal{B}$ is the collection of the balls of $\mathcal{B}$ with centers in A_1, then $\sup\{r(B) : B \in \mathcal{B}_1\} \le M/2$.

We may continue the construction in the same way, starting from A_1 and $\mathcal{B}_1$, so that we get sequences $\{A_n\}, \{\mathcal{C}_n\}, \{\mathcal{B}_n\}$, such that $\mathcal{B}_0 = \mathcal{B}$, $A_0 = A$. The collection of balls $\mathcal{C}_n \subset \mathcal{B}_{n-1} \subset \mathcal{B}$ has the property that all balls of $\mathcal{C}_n$ have centers in A_{n-1} and the balls of $\frac{1}{4}\mathcal{C}_n$ are disjoint. $A_n = A_{n-1} \setminus (\bigcup\limits_{B \in \mathcal{C}_n} B)$ and $\mathcal{B}_n = \{B = B(x,r) \in \mathcal{B} : x \in A_n\}$. Due to the construction we see that if $B \in \mathcal{C}_n$ then $M/2^n < r(B) \le M/2^{n-1}$. Also, $\sup\{r(B) : B \in \mathcal{B}_n\} \le M/2^n$. This last property implies that if $x \in A$ is the center of a ball (of $\mathcal{B}$) of radius greater than $M/2^n$ then $x \in \bigcup\limits_{k=1}^{n} \bigcup\limits_{B \in \mathcal{C}_k} B$. As every $x \in A$ is a center of some ball of $\mathcal{B}$, we see that

$$A \subset \bigcup_{k=1}^{\infty} \bigcup_{B \in \mathcal{C}_k} B.$$

49

To prove the first statement it is enough to show that $\widetilde{\mathcal{B}} = \bigcup_{k=1}^{\infty} \mathcal{C}_k$, satisfies the condition that no point is contained in more than $P(n)$ of balls of $\widetilde{\mathcal{B}}$.

Consider a point $x \in \bigcup_{B \in \widetilde{\mathcal{B}}} B$. Let $x \in B_1 = B(x_1, r_1) \in \mathcal{C}_k$ and $x \in B_2 = B(x_2, r_2) \in \mathcal{C}_m$. Without loss of generality we may assume $k < m$. Then in the triangle $\triangle x x_1 x_2$ the sides which contain x are less than the corresponding radii, while $|x_1 x_2|$ is greater than the greater of the radii, i.e. the side $x_1 x_2$ is the greatest side of the triangle. This implies that the angle at x is the greatest angle of the triangle, as the opposite to the greatest side, so it is greater than $60°$.

Exercise 7.4. There is a number $R(n)$, such that if a bunch of rays has the origin at the same point and the angle between each pair is at least $60°$, then there are no more than $R(n)$ rays. (Hint: consider the points of intersection of the rays with the unit sphere with center at the origin.)

By the exercise the point x can belong to the balls of no more than $R(n)$ different collections $\mathcal{C}_k$. To how many balls of the same collection $\mathcal{C}_k$ can x belong?

Consider a ball $B(x, r)$ of radius $r = M/2^{k-2}$ with center in x. It has to contain all the balls of $\mathcal{C}_k$, which contain x. Consider the balls in $\frac{1}{4}\mathcal{C}_k$ which are contained in $B(x, r)$. They are disjoint balls of radius at least $M/2^{k+2}$ and so (by counting volume) there are no more than 16^n balls of $\frac{1}{4}\mathcal{C}_k$ in $B(x, r)$. So there are no more than 16^n balls of $\mathcal{C}_k$ which contains x.

So, x belongs to the balls from no more than $R(n)$ collections $\mathcal{C}_k$, to no more than 16^n in each of the collections. Altogether it belongs to no more than $P(n) = R(n)16^n$ balls of $\widetilde{\mathcal{B}}$. Which proves the first part of the theorem.

Break

Remark 7.3. Observe that the first part of the theorem can be reformulated as $\sum_{B \in \widetilde{\mathcal{B}}} \chi_B(x) \leq P(n)$ for all $x \in \mathbb{R}^n$.

To prove the second part of the theorem, first, well-order $\widetilde{\mathcal{B}}$ in the following way: The balls within $\mathcal{C}_k$ should be well-ordered in an arbitrary way, but all the balls from $\mathcal{C}_k$ are less than all the balls in $\mathcal{C}_m$, whenever $k < m$.

Exercise 7.5. Show that this is a well-order relation.

Now, we will show that each ball $B \in \widetilde{\mathcal{B}}$ intersects no more than $Q(n) - 1$ previous balls. Consider $2B$. As the radius of each previous ball $\widetilde{B}$ is at least $\frac{1}{2}r(B)$, we know that $2B \cap \widetilde{B}$ contains at least a ball of radius $\frac{1}{2}r(B)$. If we denote the volume of the unit ball by v, we can write the following estimate:

$$v \left(\frac{1}{2}r(B) \right)^n \#\{\widetilde{B} : \widetilde{B} \cup B \neq \emptyset, \widetilde{B} \in \mathcal{B}, \widetilde{B} > B\}$$

$$\leq \int \sum_{\widetilde{B} \cup B \neq \emptyset; \widetilde{B} \in \widetilde{\mathcal{B}}, \widetilde{B} > B} \chi_{2B \cap \widetilde{B}}$$

$$= \int \chi_{2B} \sum_{\widetilde{B} \cup B \neq \emptyset; \widetilde{B} \in \widetilde{\mathcal{B}}} \chi_{\widetilde{B}} \leq \int \chi_{2B} P(n) \leq v(2r(B))^n.$$

By canceling out the same terms from the left and right sides of the estimate we see that B intersects no more than $Q(n) - 1 = 4^n P(n)$ previous balls.

It remains to use transitive induction[3] to show that if every ball intersects no more than $Q(n)$ previous balls then the collection can be divided in $Q(n)$ parts, such that each part contains only disjoint balls: Take the least element B which we can not place in some part. All the previous

---

[3]Transitive induction is a generalization of the induction principle from $\mathbb{N}$ to a well-ordered set O. If we can show that for each $\omega \in O$ a statement $P(\omega)$ holds for ω as soon as it holds for each $\widetilde{\omega} \prec \omega$, then the statement P holds for all elements of O.

51

balls are somehow divided between the $Q(n)$ parts of the collection. Only at most $Q(n) - 1$ parts contain balls intersecting B, so we can place also B in such a way that all parts contain disjoint balls. This contradiction shows that all the balls can be placed in one of the $Q(n)$ parts. $\square$

Vitali's covering theorem

Definition 7.4. A measure μ is called a Radon measure if

1. $\mu(K) < \infty$ for each compact K;

2. $\mu(V) = \sup\{\mu(K) : K \subset V\}$ for all open V, where K runs over the compact sets;

3. $\mu(A) = \inf(\mu(V) : A \subset V\}$ where V runs over the open sets.

Definition 7.5. We call the measure ν a *restriction* of the measure μ to the μ-measurable set K if for each μ-measurable set A we have $\nu(A) = \mu(A \cap K)$. This is denoted $\nu = \mu|_K$.

Exercise 7.6. Show that the restriction of $\mathcal{H}_\alpha$ to each compact set K of finite $\mathcal{H}_\alpha$-measure is a Radon measure.

Theorem 7.6. *Let μ be a Radon measure on $\mathbb{R}^n$, $A \subset \mathbb{R}^n$ and $\mathcal{B}$ be a family of closed balls such that each point of A is the center of arbitrarily small balls of $\mathcal{B}$. Then there are disjoint balls of $B_j \in \mathcal{B}$ such that $\mu(A \setminus (\cup B_j)) = 0$.*

Proof. Assume first that $0 < \mu(A) < \infty$. Choose an open set $V \supset A$ such that $\mu(V) < \left(1 + \dfrac{1}{4Q(n)}\right)\mu(A)$. Remove from $\mathcal{B}$ all balls except for those which are subsets of V. The condition that each point of A is center of arbitrarily small ball from $\mathcal{B}$ remains true.

By Besicovitch's covering theorem, we can find $\{\widetilde{\mathcal{B}}_k\}_{k=1}^{Q(n)}$ such that

$$A \subset \bigcup_{k=1}^{Q(n)} \bigcup_{B \in \widetilde{\mathcal{B}}_k} B$$ and the above properties hold.

We see than that $\mu(A) \leq \sum\limits_{k=1}^{Q(n)} \sum\limits_{B \in \widetilde{\mathcal{B}_k}} \mu(B)$, and so for some k we have $\mu(A) \leq Q(n) \sum\limits_{B \in \widetilde{\mathcal{B}_k}} \mu(B)$. We may pick a finite collection of (disjoint) balls $\{B_j\}_{j=1}^{m_1} \subset \widetilde{\mathcal{B}_k}$ so that $\mu(A) \leq 2Q(n) \sum\limits_{j=1}^{m_1} \mu(B_j)$.

Consider $A_1 = A \setminus (\bigcup\limits_{j=1}^{m_1} B_j)$. We see that

$$\mu(A_1) \leq \mu(V \setminus (\bigcup_{j=1}^{m_1} B_j)) \leq \left(1 - \frac{1}{4Q(n)}\right) \mu(A).$$

As $\bigcup\limits_{j=1}^{m_1} B_j$ is closed, we may remove from $\mathcal{B}$ all balls which intersect $\bigcup\limits_{j=1}^{m_1} B_j$, and still every point of A_1 will be the center of a ball of $\mathcal{B}$ of arbitrarily small radius.

If we continue the construction we will get a sequence $\{B_j\}$ of disjoint balls of $\mathcal{B}$ such that $A_p = A \setminus (\bigcup\limits_{j=1}^{m_p} B_j)$ and $\mu(A_p) \leq \left(1 - \frac{1}{4Q(n)}\right)^p \mu(A)$. Making infinitely many steps (or rather choosing a maximal chain of all possible sequences of the steps) we see that $\mu(A \setminus (\bigcup\limits_{j=1}^{\infty} B_j)) = 0$. $\qquad\square$

Exercise 7.7. Complete the proof for $\mu(A) = \infty$. (Hint: Divide A into pieces of finite measure (and possibly a residue of measure zero), so that the set $\mathcal{B}$ would include the corresponding sub-collections.)

Exercise* 7.8. Show that $\mathcal{H}_\alpha$ on $\mathbb{R}^n$ for $\alpha < n$ is not a Radon measure.

Lecture 8
Measure and integral

In order to compare measures we start by introducing an integral with respect to a measure. It goes pretty much as one expects after all the definitions of the integral to which one was exposed in the undergraduate courses. Yet, the first year calculus definition often has a flaw: We talk about the limit, when the length of the maximal interval in the decomposition tends to 0. It is possible to make the definition this way, but it is rarely done carefully.

Exercise* 8.1. Give a careful ϵ, δ-definition of the Riemann integral.

While it is possible to use pure ϵ, δ to obtain the definition of the integral, but it is interesting to see if the ϵ, δ definition can be expressed in the language of sequences[1] (as can be done, for example, with the definition of continuity).

Definition 8.1. A partially ordered set $(A, \prec)$ is called a *directed* set if for any $a, b \in A$ there exists $c \in A$ such that $a \prec c$ and $b \prec c$. A *net* is a function from a directed set to a topological space. If $x(a)$ is a net in a topological space X, we say that the net has a *limit* $x = \lim x(a)$ if for each neighborhood U of x there exists $\alpha(U) \in A$ such that $x(b) \in U$ for every $b \succ \alpha$.

Exercise* 8.2. Show that the partitions of an interval $[p, q]$ form a directed set (under the relation of subpartition), and the Riemann sums of a continuous f corresponding to them is a convergent net, with limit $\int_p^q f(x)dx$.

Back to the measures:

Definition 8.2. Given a measure μ, a function is called μ-*measurable* if for any $\alpha, \beta \in \mathbb{R}$ the set $\{x : \alpha \leq f(x) < \beta\}$ is μ-measurable.

---

[1]There is a French way to do it with ultra-filters, but we do it in a non-French fashion with nets.

For a μ-measurable non-negative function f we consider the collection $\mathcal{I}$ of sums

$$S(f, \{\alpha_k\}_{k=0}^\infty) = \sum_{k=0}^\infty \alpha_k \mu(\{x : \alpha_k \le f(x) < \alpha_{k+1}\}),$$

for all sequences $0 = \alpha_0 < \alpha_1 < \ldots$ that are increasing to infinity. The collection $\mathcal{I}_\delta \subset \mathcal{I}$ consists of those sequences which, in addition, satisfy $\sup(\alpha_{k+1} - \alpha_k) \le \delta$. It is clear that the collection $\mathcal{I}_\delta$ is increasing with δ.

Exercise 8.3. Show that when $\sup(\mathcal{I}_\delta) < \infty$ for some $\delta > 0$ it is so for all $\delta > 0$, and $\sup(\mathcal{I}_\delta) - \inf(\mathcal{I}_\delta) \to 0$ when $\delta \to 0$. (Hint: Observe that $\sup(\mathcal{I}_\delta)$ is increasing with δ as so is $\mathcal{I}_\delta$, but if one adds points to a sequence the corresponding sum increases.)

Exercise 8.4. Show that when $\sup(\mathcal{I}_\delta) = \infty$ for some $\delta > 0$ then $\inf(\mathcal{I}_\delta) \to \infty$ when $\delta \to 0$.

Definition 8.3. For a positive μ-measurable function f we call

$$\int f(x) d\mu(x) = \lim_{\sup(a_{k+1}-a_k)\to 0} \sum_{k=0}^\infty \alpha_k \mu(\{x : \alpha_k \le f(x) < \alpha_{k+1}\})$$

the *integral* of f with respect to the measure μ. The integral with respect to the Lebesgue measure is called the *Lebesgue integral*.

Remark 8.4. Notice that the integral of a positive function can be infinite.

Exercise 8.5. Introduce a definition of the integral with respect to a measure μ via net limits.

Exercise 8.6. Show that if $0 \le f_1 \le f_2$ then $\int f_1(x) d\mu(x) \le \int f_2(x) d\mu(x)$.

Definition 8.5. A μ-measurable function f is called μ-*integrable* if $\int |f(x)| d\mu(x) < \infty$. Then $\int f(x) d\mu(x) = \int f_+(x) d\mu(x) - \int f_-(x) d\mu(x)$, is the *integral* of f with respect to μ (here $f_+ = \sup(f, 0), f_- = -\inf(f, 0)$).

Exercise 8.7. Show that if f is μ-measurable, so are $f_+, f_-, |f|$.

56

Observe, that given a measure μ and a non-negative integrable function f we can introduce a measure ν, by setting

$$\nu(A) = \int_A f(x)d\mu(x) = \int \chi_A(x)f(x)d\mu(x).$$

Exercise 8.8. Show that if f is an integrable function and A is a measurable set (both in respect to the measure μ) then $\chi_A(x)f(x)$ is measurable with respect to the measure μ.

Now, let us have two measures: μ and ν.

Definition 8.6. We say that the measure ν is *absolutely continuous* with respect to the measure μ if every set measurable with respect to ν is measurable with respect to μ, and if $\mu(A) = 0$ then $\nu(A) = 0$ (the converse is not assumed and is not necessarily true).

Exercise 8.9. Show that for a non-negative function f, integrable with respect to a measure μ, the measure $\nu = f\mu$ is absolutely continuous with respect to the measure μ.

This is a very practical definition to verify, and its connection with the previous discussion follows from the following:

Theorem 8.7 (Radon-Nikodym decomposition theorem). *Let ν and μ be two finite measures (i.e. the measure of the whole space is finite). If ν is absolutely continuous with respect to μ, then there exists a μ-measurable non-negative function f such that for each ν-measurable set A the function $\chi_A f$ is μ-measurable and $\nu(A) = \int_A f(x)d\mu(x)$.*

Proof. Consider the collection $\mathcal{F}$ of all non-negative ν-measurable[2] functions h such that for each ν-measurable A we have $\int_A h(x)d\mu(x) \leq \nu(A)$. The family $\mathcal{F}$ is non-empty, as it contains the zero-function.

Consider a sequence of functions $\{h_n\} \subset \mathcal{F}$ such that $\int_X h_n(x)d\mu(x) \to \sup_{h \in \mathcal{F}} \int_X h(x)d\mu(x)$. Introduce $g_n(x) = \max\{h_k(x) : k \leq n\}$.

---

[2] As each ν-measurable set is μ-measurable, each ν-measurable function is μ measurable.

Exercise 8.10. Show that the maximum of two ν-measurable functions is ν-measurable. Hint: It is enough to show that the set $\{x : f_1(x) > f_2(x)\}$ is ν-measurable, and to do that consider for every $r \in \mathbb{Q}$ the set $\{x : f_1(x) > r > f_2(x)\}$.

Let $A_1 = \{h_1 > h_2\}$ and $A_2 = \{h_1 \leq h_2\}$. Then for each ν-measurable A we have

$$
\begin{aligned}
\nu(A) &= \nu(A \cap A_1) + \nu(A \cap A_2) \\
&\geq \int_{A \cap A_1} h_1(x)d\mu(x) + \int_{A \cap A_2} h_2(x)d\mu(x) \\
&= \int_{A \cap A_1} g_2(x)d\mu(x) + \int_{A \cap A_2} g_2(x)d\mu(x) = \int_A g_2(x)d\mu(x),
\end{aligned}
$$

i.e. $g_2 \in \mathcal{F}$. Similarly we see that $\{g_n\} \subset \mathcal{F}$.

Now, the sequence of functions (g_n) is increasing, so we may define $g(x) = \lim g_n(x)$.

Exercise 8.11. Show that when f is a limit of increasing sequence of positive μ-measurable functions f_n then f is μ-measurable and $\lim \int f_n(x)d\mu(x) = \int f(x)d\mu(x)$. Hint: Use the definition of the integral, and the fact that the measure μ is finite.

For each ν-measurable set A we have $\nu(A) \geq \lim \int_A g_n(x)d\mu(x) = \int_A g(x)d\mu(x)$. Thus $g \in \mathcal{F}$ and $\sup\{\int_X h(x)d\mu(x) : h \in \mathcal{F}\}$ is attained on g.

If $\int_x g(x)d\mu(x) = \nu(X)$ then for every ν-measurable set A we have $\int_A g(x)d\mu(x) \leq \nu(A)$ and $\int_{X \setminus A} g(x)d\mu(x) \leq \nu(X \setminus A)$. In order for both inequalities hold they have to be equalities and we have shown that $\nu = f\mu$.

Break

Assume $\int_x g(x)d\mu(x) < \nu(X)$. We can choose an $\varepsilon > 0$ such that $\int_X (g(x) + \varepsilon)d\mu(x) < \nu(X)$. We claim[3] that there exists a ν-measurable

---

[3]The following argument is an incorporated proof of the Hahn decomposition theorem: Each signed measure can be presented as a difference of two positive measures.

set $P \subset X$, $\nu(P) > 0$, such that for all ν-measurable subsets $B \subset P$ we have $\int_B (g(x) + \varepsilon)d\mu(x) \leq \nu(B)$.

Indeed, start from $P_0 = X$ and consider

$$t_0 = \sup \left\{ \int_B (g(x) + \varepsilon)d\mu(x) - \nu(B) : B \subset P_0 \right\}.$$

If $t_0 \leq 0$ then P_0 satisfies the required condition and we are done. Choose B_0 such that $\int_{B_0} (g(x) + \varepsilon)d\mu(x) - \nu(B) \geq \frac{1}{2}t_0$. Take now $P_1 = P_0 \setminus B_0$. Unless at some point t_n becomes non-positive (in which case we are done), we obtain an infinite sequence of sets $\{P_n\}$ and $\{B_n\}$, where $\{P_n\}$ is decreasing so we may take $P = \cap P_n = X \setminus \bigcup B_n$, and B_n are disjoint.

For every B_n we have $\int_{B_n} (g(x) + \varepsilon)d\mu(x) > \nu(B_n)$, so

$$\int_{\cup_n B_n} (g(x) + \varepsilon)d\mu(x) > \nu(\cup_n B_n),$$

and thus $\cup B_n \neq X$ (as $\int_X (g(x) + \varepsilon)d\mu(x) \leq \nu(X)$). As $P = X \setminus (\cap B_n)$, it is not empty. Even more so, as

$$\int_P (g(x) + \varepsilon)d\mu(x) - \nu(P) < \int_X (g(x) + \varepsilon)d\mu(x) - \nu(X) < 0,$$

we see that $\nu(P) > 0$.

Now, we should prove that for each ν-measurable $B \subset P$ we have $\int_B (g(x) + \varepsilon)d\mu(x) \leq \nu(B)$.

As $\frac{1}{2}t_n \leq \int_{B_n} (g(x) + \varepsilon)d\mu(x)$ and the sets B_n are disjoint we see that

$$\sum t_n \leq 2 \int_{\cup B_n} (g(x)+\varepsilon)d\mu(x) \leq 2 \int_X (g(x)+\varepsilon)d\mu(x) \leq 2\nu(X)+\varepsilon\mu(X) <$$

∞, so $t_n \to 0$. This means that

$$\sup \left\{ \int_B (g(x) + \varepsilon)d\mu(x) - \nu(B) : B \subset P \right\}$$

As the signed measures are outside the scope of this book, the author has adjusted the proof to avoid mentioning them.

$$\leq \limsup \left\{ \int_B (g(x) + \varepsilon) d\mu(x) - \nu(B) : B \subset P_n \right\} = 0.$$

It remains to observe that for each ν-measurable set A we have

$$\int_A (g(x) + \varepsilon \chi_P(x)) d\mu(x)$$
$$= \int_{A \setminus P} g(x) d\mu(x) + \int_{P \cap A} (g(x) + \varepsilon) d\mu(x)$$
$$\leq \nu(A \setminus P) + \nu(A \cap P) = \nu(A),$$

thus $g + \varepsilon \chi_P \in \mathcal{F}$. If $\mu(P) = 0$ then $\nu(P) = 0$ which is not the case, so

$$\int_A (g(x) + \varepsilon \chi_P(x)) d\mu(x)$$
$$> \int_A g(x) d\mu(x) = \sup\{ \int_X h(x) d\mu(x) : h \in \mathcal{F} \},$$

and this contradiction completes the proof. $\qquad\qquad\square$

The opposite notion to that of the absolute continuity of measures is that of mutual singularity.

Definition 8.8. Measures ν and μ are called mutually *singular* if there exists a μ- and ν-measurable set A that $\mu(A) = 0$ and $\nu(X \setminus A) = 0$.

A complement to the Radon-Nikodym theorem is the Lebesgue decomposition theorem which says that given two finite measures ν and μ defined on the same σ-algebra we can represent ν as a sum $\nu = \nu_1 + \nu_2$, where ν_1 is absolutely continuous with respect to μ and ν_2 is mutually singular to μ.

Exercise 8.12. Prove the Lebesgue decomposition theorem. Hint: Employ ideas from the proof of Hahn decomposition theorem.

Combining the two theorems we see that, given two finite measures μ and ν, we can represent $\nu = f\mu + \nu_s$, where ν_s is singular to μ.

Both the Radon-Nikodym and the Lebesgue theorems can be proven explicitly using covering theorems. One can do so by looking at the density function (this we do in the next lecture), but let us first see a way of estimating the Hausdorff measure of a set from below.

Theorem 8.9. *If for a compact set K there exists a nonzero finite measure such that $\mu(X \setminus K) = 0$ (i.e. the measure is supported on K) and $\mu(B(x,r)) \leq cr^\alpha$ for every x and $r < r_0$, where r_0 and c are constants. Then $\mathcal{H}_\alpha(K) > 0$.*

Proof. Consider a covering of K by balls $\{B_j\}$ of radii less than r_0. We see that

$$\mu(K) = \mu(\cup B_j) \leq \sum \mu(B_j) \leq c \sum r^\alpha(B_j).$$

This implies that $\mathcal{S}_\alpha^\delta \geq \mu(K)/c$ for all $0 < \delta < r_0$. So,

$$\mathcal{S}_\alpha(K) \geq \mu(K)/c > 0.$$

$\square$

Exercise 8.13. Using the theorem, find the Hausdorff dimension of a unit interval, a square and a cube.

Exercise 8.14. Find the Hausdorff dimension of $\frac{1}{3}$-Cantor set. Hint: We already have an estimate from above.

Lecture 9
Application of the covering theorems

Recall that to estimate from below the Hausdorff measure/dimension of a self-similar set (or any other set) we need to construct a nice measure on it. Let a self-similar set K correspond to a collection of similitudes $\{\varphi_k\}_{k=1}^m$. We have guessed before that the greatest dimension of Hausdorff measure which is positive on K is α such that $\sum r_k^\alpha = 1$, where r_k is the coefficient of the similitude ϕ_k. We have seen that $\mathcal{H}_{\alpha'}(K) = 0$ for all $\alpha' > \alpha$ and that $\mathcal{H}_\alpha(K) < C < \infty$.

However, we need one more condition, to avoid the following situation: Let ϕ_1 be the contraction by $\frac{3}{4}$ towards 0, without rotation, and ϕ_2 be the contraction by $\frac{3}{4}$ towards 1, with no rotation. Then the corresponding self-similar set is the interval $[0,1]$ which has dimension 1 and not the expected $\dfrac{\log 2}{\log 4 - \log 3} \approx 2 \cdot 4$. The problem to avoid is that the sets in the union $K = \cup \phi_k(K)$ do not overlap too much.

Definition 9.1. We say that the *open set condition* holds for the similitudes $\{\phi_k\}$ if there exists a bounded open set V such that $\cup \phi_k(V) \subset V$ and the sets in the union are disjoint.

Theorem 9.2. *Let the collection of similitudes $\{\phi_k\}$ satisfy the open set condition. Then the corresponding self-similar set $K = \cup \phi_k(K)$ satisfies $\mathcal{H}_\alpha(K) > 0$, where $\sum r_k^\alpha = 1$, where the r_k are the coefficients of the similitudes.*

Again, to prove the theorem it is enough to construct a measure $\mu(K) > 0$, $\mu(\mathbb{R}^d \setminus K) = 0$, such that for each ball of sufficiently small radius $\mu(B) \le c(r(B))^\alpha$. To do the construction we need the following.

Theorem 9.3 (Riesz representation theorem[1]). *If $\ell : C_0(\mathbb{R}^n) \mapsto \mathbb{R}$ is a linear functional (on the space of continuous functions which vanish at*

---

[1] The Riesz representation theorem is valid on a locally compact metric space

infinity), and has the property that on positive functions it is positive, then there exists a unique finite Radon measure μ such that $\ell(\phi) = \int \phi d\mu$.

Exercise 9.1. Prove the existence part of the Riesz representation theorem. Hint. Define μ on compact sets as $\inf\{\ell(\phi) : 0 \leq \phi \leq 1, \phi|_K \equiv 1\}$. Then extend the definition to open sets and use inner and outer measures as we did in Lecture 5.

Exercise 9.2. Show uniqueness in the Riesz representation theorem. Hint: Assume there are two measures which give the same functional. Use the Lebesgue decomposition theorem, and don't forget what it means that the measures are Radon (as well as the fact that given two disjoint compacts we always have a continuous function which is 1 on one and 0 on the other).

Proof of Theorem 9.2. Let $\sup\{r_k\} = R$ and $\inf\{r_k\} = r$. $r, R < 1$.
Consider the map $\Phi : C_0(X) \to C_0(X)$ such that

$$\Phi(f) = \sum r_k^\alpha f \circ \phi_k.$$

Notice that as $\sum r_j^\alpha = 1$ the map $\Phi(f)$ has norm not greater than 1 on $C_0(X)$ (with the sup-norm). Also to each function $f \in C_0(X)$ there corresponds a function that increases from zero, given by $\varepsilon_f(\delta) = \sup\{|f(x) - f(y)| : |x - y| \leq \delta\}$. The map Φ satisfies $\varepsilon_{\Phi(f)}(\delta) \leq \varepsilon_f(r\delta)$. Thus $\Phi^{\circ n}(f)$ converges pointwise to a constant for every f. Let $\ell(f) = \lim_{n \to \infty} \Phi^{\circ n}(f)$.

Exercise 9.3. Check that ℓ is a linear functional on $C_0(X)$.

By the Riesz representation theorem there exists a Radon measure μ such that $\ell(f) = \int f d\mu$. Observe that $\int f d\mu = \ell(f) = \ell(\Phi(f)) = \sum r_k^\alpha \ell(f \circ \phi_k) = \sum r_k^\alpha \int f \circ \phi_k d\mu = \sum r_k^\alpha \int f d\mu \circ \phi_k^{-1} = \int f d(\sum r_k^\alpha \mu \circ \phi_k^{-1})$. By uniqueness in the Riesz representation theorem this gives $\mu = \sum r_k^\alpha \mu \circ \phi_k^{-1}$, where $\mu \circ \phi_k^{-1}(A) = \mu(\phi_k^{-1}(A))$.

Exercise 9.4. Consider what happens with a function f with support outside of K (i.e. which vanish on K). Show that $\ell(f) = 0$, thus proving that $\mu(\mathbb{R}^d \setminus K) = 0$.

64

Consider an arbitrary ball $B(x, \rho)$, $\rho < R$. We want to show that $\mu(B) \leq c\rho^\alpha$.

The similitudes satisfy the open set condition. This means that we have $V \supset \cup \phi_k(V)$ and also that $\overline{V} \supset \cup \phi_k(\overline{V})$. If we construct K starting from $\overline{V}$ we see that $K \subset \overline{V}$.

Let us start by considering (finite) iterations of the set V under different compositions of ϕ_k. We can introduce a partial order on the compositions where we say that $\phi_{k_1} \circ \ldots \circ \phi_{k_n} \prec \phi_{k_1} \circ \ldots \circ \phi_{k_m}$ if the sequence $(k_j)_{j=1}^m$ consists of the first m elements of the sequence $(k_j)_{j=1}^n$. In this case (as $\phi_k(\overline{V}) \subset \overline{V}$ for all k) we know that $\phi_{k_1} \circ \ldots \circ \phi_{k_n}(\overline{V}) \subset \phi_{k_1} \circ \ldots \circ \phi_{k_m}(\overline{V})$.

Exercise 9.5. Show that if $\phi_{k_1} \circ \ldots \circ \phi_{k_n}$ and $\phi_{k_1} \circ \ldots \circ \phi_{k_m}$ are incomparable than $\phi_{k_1} \circ \ldots \circ \phi_{k_n}(V)$ and $\phi_{k_1} \circ \ldots \circ \phi_{k_m}(V)$ are disjoint.

We will consider $\widetilde{S} = \{(k_1, \ldots, k_n) : r\rho < r_{k_1} \ldots r_{k_n} \leq \rho\}$, and S - the set of maximal elements in $\widetilde{S}$.

Observe, first, that as $r_{k_1} \ldots r_{k_n} \leq R^n$ and $R < 1$ the length of every sequence of $\widetilde{S}$ can be no more than $\dfrac{\log(\rho)}{\log(R)} + 1$. Then, each sequence is comparable (i.e. is less, equal or greater) with some sequence of S and no two sequences in S are comparable (as they all are maximal).

Exercise 9.6. Show that $\mu = \sum\limits_{(k_1, \ldots, k_n) \in S} r_{k_1}^\alpha \ldots r_{k_n}^\alpha \mu \circ \phi_{k_n}^{-1} \circ \ldots \circ \phi_{k_1}^{-1}$.
Hint: Use that $\mu = \sum r_k^\alpha \mu \circ \phi_k^{-1}$ and the properties of S.

Let $\mathcal{I} \subset S$ consist of the sequences $(k_1, \ldots, k_n)$ such that $B(x, \rho)$ intersects $\phi_{k_1} \circ \ldots \circ \phi_{k_n}(\overline{V})$. As

$$\text{diam}(\phi_{k_1} \circ \ldots \circ \phi_{k_n}(\overline{V})) = r_{k_1} \ldots r_{k_n} \text{diam}(V) \leq \rho \text{diam}(V),$$

we see that $B(x, (\text{diam}(V) + 1)\rho)$ contains all $\phi_{k_1} \circ \ldots \circ \phi_{k_n}(V)$, for $(k_1, \ldots, k_n) \in \mathcal{I}$. On the other hand

$$\text{Vol}(\phi_{k_1} \circ \ldots \circ \phi_{k_n}(V)) = r_{k_1}^d \ldots r_{k_n}^d \cdot \text{Vol}(V)$$
$$\geq r\rho^d \cdot \text{Vol}(V) \geq C_{d,r,V}^{-1} \cdot \text{Vol}(B(x, (\text{diam}(V) + 1)\rho)).$$

As all involved sets $\phi_{k_1} \circ \ldots \circ \phi_{k_n}(V)$ are disjoint, this implies that $\#\mathcal{I} < C_{d,r,V}$.

Now, if $\phi_{k_1} \circ \ldots \circ \phi_{k_n}(\overline{V}) \cap B(x, \rho) = \emptyset$, then

$$\phi_{k_n}^{-1} \circ \ldots \circ \phi_{k_1}^{-1}(B(x, \rho)) \cap \overline{V} = \emptyset$$

and so $\mu \circ \phi_{k_n}^{-1} \circ \ldots \circ \phi_{k_1}^{-1}(B(x, \rho)) = 0$. Thus,

$$
\begin{aligned}
\mu(B(x, \rho)) &= \sum_{(k_1, \ldots, k_n) \in \mathcal{I}} r_{k_1}^{\alpha} \ldots r_{k_n}^{\alpha} \mu \circ \phi_{k_n}^{-1} \circ \ldots \circ \phi_{k_1}^{-1}(B(x, \rho)) \\
&\leq \sum_{(k_1, \ldots, k_n) \in \mathcal{I}} r_{k_1}^{\alpha} \ldots r_{k_n}^{\alpha} \mu(\mathbb{R}^d) \\
&\leq C_{d,r,V} \rho^{\alpha},
\end{aligned}
$$

which is what we wanted to prove. $\square$

Observe that the definition of Hausdorff measure implies that if $\mathcal{H}_{\alpha}(K) < \infty$ then $\mathcal{H}_{\alpha'}(K) = 0$ for all $\alpha' > \alpha$, and if $\mathcal{H}_{\alpha}(K) > 0$ then $\mathcal{H}_{\alpha'}(K) = \infty$ for all $\alpha' < \alpha$. Thus, if one considers $\mathcal{H}_{\alpha}(K)$ as a function of α, the function is first infinite and then after a given value becomes 0. The point of the break is called the *Hausdorff dimension* of the set K.

While every set has some Hausdorff dimension, if we have $dim_H(K) = \alpha$ this doesn't mean that $0 < \mathcal{H}_{\alpha}(K) < \infty$.

Exercise 9.7. Construct sets such that $dim_H(K) = \alpha$ but $\mathcal{H}_{\alpha}(K) = 0$ or $\mathcal{H}_{\alpha}(K) = \infty$. Hint: One can modify the Cantor construction, so that on each step one the middle interval which we throw away has different proportional length.

The theorem above has used the theorem, which we have proved the previous time, i.e. if a set supports a positive measure such that $\mu(B) \leq cr^{\alpha}(B)$ for all (or at least sufficiently small) balls B, then the set has positive α-Hausdorff measure. This means that the set has Hausdorff dimension at least α. If we are interested only in the Hausdorff dimension, a weaker theorem suffices.

Theorem 9.4. *If K is a closed set and there exists a nonzero measure μ such that $\mu(X \setminus K) = 0$ and $\int \dfrac{d\mu(x)d\mu(y)}{d(x,y)^{\alpha}} < \infty$ then the Hausdorff dimension of the set K is at least α.*

66

Proof. Let $\displaystyle\int \frac{d\mu(x)d\mu(y)}{d(x,y)^\alpha} = C.$

Assume, that for every $N > 0$ there exists a closed ball $B = B(x,r)$, such that $\mu(B) > NCr^\alpha$. Then, for every $x \in B$ we have $\mu(B(x,2r)) > \mu(B) \geq NCr^\alpha$.

Thus,

$$\int\limits_{x\in B, y\in X} \frac{1}{d(x,y)^\alpha} d\mu(y)d\mu(x) \geq \int\limits_{x\in B} \left(\int\limits_{B(x,2r)} \frac{1}{d(x,y)^\alpha} d\mu(y) \right) d\mu(x)$$

$$\geq \mu(B)(3r)^{-\alpha}\mu(y \in B(x,2r)) \geq NC3^{-\alpha}\mu(B).$$

Now, we denote by K_N the set of points x for which there exists a ball $B(x,r)$, $r < r_0$ for which $\mu(B) > Nr^\alpha(B)$.

Exercise 9.8. Show that K_N is open.

Consider the collection $\mathcal{B}$ of such a balls. By the Besicovitch covering theorem we can consider $\{B_j\}$ which still covers K_N, but with no point covered more than $P(n)$ times.

For this covering

$$\mu(\cup B_j) \leq \sum \mu(B_j) \leq \frac{3^\alpha}{NC} \sum \int_{x\in B_j} \frac{1}{d(x,y)^\alpha} d\mu(y)d\mu(x)$$

$$\leq \frac{P(n)3^\alpha}{CN} \int \frac{1}{d(x,y)^\alpha} d\mu(y)d\mu(x) \leq \frac{P(n)3^\alpha}{N}.$$

This means that if we choose $N > P(n)3^\alpha\mu(X)$, then $\mu(K \setminus K_N) \geq \mu(K \setminus (\cup B_j)) > 0$. Consider $\nu = \mu|_{X\setminus K_N}$. The measure ν is less than μ and for each point of its support all balls with center there and of radius less than r_0 satisfy $\nu(B) \leq NCr^\alpha(B)$. Now we can use the theorem from the previous lecture to see that $\mathcal{H}_\alpha(K \setminus K_N) > 0$. □

Exercise 9.9. The theorem we had in the previous lecture required all balls (not only those with the centers in its support) to satisfy $\nu(B) \leq cr^\alpha(B)$. Show that the required modification of it is still correct.

Break

Definition 9.5. The upper and lower derivative of the measure ν with respect to the measure μ are defined everywhere by

$$\overline{D(\nu,\mu)}(x) = \limsup_{r \to 0}\{\nu(B(x,r))/\mu(B(x,r))\},$$

$$\underline{D(\nu,\mu)}(x) = \liminf_{r \to 0}\{\nu(B(x,r))/\mu(B(x,r))\},$$

and when they coincide we talk about the density function

$$D(\nu,\mu)(x) = \lim_{r \to 0}\{\nu(B(x,r))/\mu(B(x,r))\}.$$

Here by $B(x,r)$ we mean the closed ball with center at x and radius r, and we regard $\frac{0}{0} = +\infty$.

Theorem 9.6. *Let μ and ν be Radon measures. Then the upper and lower derivatives are Borel-measurable (and thus μ-measurable), $D(\nu,\mu)$ is defined μ-almost everywhere, and for each Borel set E we have $\int_E D(\nu,\mu)(x)d\mu(x) \le \nu(E)$ (the equality holds for all E if ν is absolutely continuous with respect to μ).*

Proof. First, let us introduce $\overline{D_n(\nu,\mu)}(x) = \sup\limits_{r < \frac{1}{n}}\{\nu(B(x,r))/\mu(B(x,r))\}$.

We claim that the set $\{x : \overline{D_n(\nu,\mu)}(x) > a\}$ is open for every a. Indeed, let r be such that $\nu(B(x,r))/\mu(B(x,r)) > a$. If $\mu(B(x,r)) = 0$ then the same is true for $B(y,\frac{1}{2}r)$, $y \in B(x,\frac{1}{2}r)$.

If $\mu(B(x,r)) > 0$ then there exists an r' such that $r < r' < \frac{1}{n}$ and $\mu(B(x,r')) \le (1+\varepsilon)\mu(B(x,r))$ (remember that the balls are closed). Thus,

$$\nu(B(y,\frac{r+r'}{2}))/\mu(B(y,\frac{r+r'}{2})) \ge \nu(B(x,r))/\mu(B(x,r'))$$

$$\ge \frac{1}{1+\varepsilon}\nu(B(x,r))/\mu(B(x,r), \text{ for } d(x,y) \le \frac{1}{2}(r'-r).$$ As we could choose ε

such that $\frac{1}{1+\varepsilon}\nu(B(x,r))/\mu(B(x,r) > a$, this means that if for some x we have $\nu(B(x,r))/\mu(B(x,r) > a$ so we do for points in some neighborhood of this x, i.e. the set $\{x : \overline{D_n(\nu,\mu)}(x) > a\}$ is open, and so Borel-measurable.

Now, $\overline{D(\nu,\mu)} = \lim_n \overline{D_n(\nu,\mu)}$, and thus is Borel-measurable, which implies μ-measurability.

Similarly we can show that the lower derivative is Borel-measurable. We need the following.

Lemma 9.7. *Let μ and ν be Radon measures, A be a Borel set and $0 < p < \infty$, then*

- *If $\overline{D(\nu,\mu)}(x) \geq a$ for all $x \in A$, then $\nu(A) \geq a\mu(A)$;*

- *If $\underline{D(\nu,\mu)}(x) \leq a$ for all $x \in A$, then $\nu(A) \leq a\mu(A)$;*

Proof of the Lemma. Take $\varepsilon > 0$. Choose an open set $U \supset A$, such that $\nu(U) \leq \nu(A) + \varepsilon$. By the assumption, $\overline{D(\nu,\mu)}(x) > a - \varepsilon$, so for each x there exists an arbitrary small ball B with center in x, such that $B \subset U$ and $\nu(B(x,r))/\mu(B(x,r)) > a - \varepsilon$. Consider $\mathcal{B}$, the collection of all such a balls. By Vitali's covering theorem we can choose a disjoint sequence of balls $\{B_j\}$ such that $\mu(A \setminus (\cup B_j)) = 0$. Now, $(a - \varepsilon)\mu(A) \leq (a-\varepsilon)\mu(\cup B_j) = (a-\varepsilon)\sum \mu(B_j) \leq \sum \nu(B_j) = \nu(\cup B_j) \leq \nu(U) \leq \nu(A)+\varepsilon$. As ε can be chosen arbitrarily small this proves the first part of the lemma. The second part can be proven similarly. $\square$

Now, observe that for each x we have $\underline{D(\nu,\mu)}(x) \leq \overline{D(\nu,\mu)}(x)$. At the same time if $\underline{D(\nu,\mu)}(x) < \overline{D(\nu,\mu)}(x)$ there exist numbers $p, q \in \mathbb{Q}$ such that $\underline{D(\nu,\mu)}(x) \leq p < q \leq \overline{D(\nu,\mu)}(x)$. Denote

$$A_{p,q} = \{x : \underline{D(\nu,\mu)}(x) \leq p < q \leq \overline{D(\nu,\mu)}(x)\}.$$

If $\mu(A_{p,q}) > 0$ then, by the lemma,

$$\nu(A_{p,q}) \leq p\mu(A_{p,q}) < q\mu(A_{p,q}) \leq \nu(A_{p,q}),$$

which is impossible. Thus $\mu(A_{p,q}) = 0$ for every couple $p < q$. But the set on which the derivative doesn't exists, i.e. $\overline{D} \neq \underline{D}$ is exactly the set $\bigcup_{p<q} A_{p,q}$. As this is a countable union of μ-null sets, its measure is zero.

Exercise 9.10. Prove that $\int_E D(\nu,\mu)(x)d\mu(x) \leq \nu(E)$ for every Borel set E. Hint: Use the definition of the integral, and Lemma 9.7.

Exercise 9.11. Show that if ν is absolutely continuous with respect to μ, then we have the equality $\int_E D(\nu, \mu)(x) d\mu(x) = \nu(E)$. Hint: If $\nu = f\mu$, consider the sets $\{f < p < q < D\}, \{f > p > q > D\}$, and use Lemma 9.7.

$\square$

Remark 9.8. One can try to replace the measure in the denominator by a power of the radius. This provides some useful information about the Hausdorff dimension of sets of positive measure, and the Borel sets on which the measure is concentrated.

Lecture 10
Hausdorff dimension and sums, part I

We have already used:

Definition 10.1. For a set A the *Hausdorff dimension* is $\dim_H(A) = \inf\{\alpha : \mathcal{H}^\alpha(A) = 0\} = \sup\{\alpha : \mathcal{H}^\alpha(A) = +\infty\}$.

Definition 10.2. The *Minkowski sum* of two sets A and B is $A + B = \{x + y : x \in A, y \in B\}$.

It is obvious that $\dim_H(A + B) \geq \max(\dim_H(A), \dim_H(B))$. Unfortunately, nothing better can be said.

In the last two lectures we look at a relatively fresh (published in 2008) result of Tom Körner. On the one hand, it illustrates the fact that the Hausdorff dimension of a set can be rather nontrivial. At the same time the argument invokes a very wide range of techniques from different fields of Analysis, so the author deems the exposure to be challenging but educational for a student.

We denote the unit circle by $\mathbb{T}$.

Theorem 10.3. *Given a non-decreasing sequence $\{\alpha_j\} \subset [0, 1)$, we can find a closed set $E \subset \mathbb{T}$ such that $\dim_H(E_j) = \alpha_j$ where $E_j = \underbrace{E + \ldots + E}_{j}$*

is the Minkowski sum of j copies of the set E.

Remark 10.4. Notice that the sequence can be very non-trivial, like being constant for a while and then jump. So the theorem shows that the sums of sets can change dimension in a strange way.

Exercise* 10.1. Show that all E_j are closed.

To prove the theorem let us consider the set $\mathcal{F}$ of all compacts in $\mathbb{T}$, metricized by the Hausdorff metric, i.e.

$$d(K_1, K_2) = \max(\sup\{d(x, K_2) : x \in K_1\}, \sup\{d(K_1, y) : y \in K_2\}).$$

71

We start by introducing a closed subset $\mathcal{K} \subset \mathcal{F}$: A set E belongs to $\mathcal{K}$ if for each j there exists a probability measure μ_j such that $\operatorname{supp}(\mu_j) \subset E_j$ and $\int \dfrac{d\mu_j(x)d\mu_j(y)}{|x-y|^{\alpha_j}} \leq K_j$, where $K_j = 6 \displaystyle\int_{\mathbb{T}^2} \dfrac{dxdy}{|x-y|^{\alpha_j}} + 1$.

First, notice that $\mathcal{K}$ is non-empty, as it contains $\mathbb{T}$. Second, observe that if $E \in \mathcal{K}$ then by the criterion introduced in the last lecture, $\dim_H(E_j) \geq \alpha_j$. Finally, let us proved that $\mathcal{K}$ is indeed closed.

Lemma 10.5. *Let $F_n \to E$ in $\mathcal{F}$ and $F_n \in \mathcal{K}$. Then $E \in \mathcal{K}$.*

Proof. We want to construct for every j a probability measure μ_j such that $\operatorname{supp}(\mu_j) \subset E_j$ and $\int \dfrac{d\mu_j(x)d\mu_j(y)}{|x-y|^{\alpha_j}} \leq K_j$.

As every μ_j can be constructed independently from the others let us fix j and concentrate on the relevant construction.

We know that for every F_n there exists ν_n such that $\operatorname{supp}(\nu_n) \subset (F_n)_j$ and $\int \dfrac{d\nu_n(x)d\nu_n(y)}{|x-y|^{\alpha_j}} \leq K_j$.

Now, recall that every finite measure corresponds to a positive linear functional on $C_0(\mathcal{R})$. Denote the functional which corresponds to ν_n as ℓ_n. What we will do now is to choose a weak*-convergent[1] subsequence of ν_n (the fact that such a sequence exists is called in Functional Analysis "the unit ball of the dual of a separable space is *-weakly compact" but we will show it now without use of the terminology of Functional Analysis).

Take a countable collection $\{\varphi_m\}$ dense in $C_0(\mathbb{R})$. Consider the sequences $\ell_n(\varphi_m)$ in n for every fixed m. As the measures are probabilistic each sequence is bounded by $\sup\{|\varphi_m|\}$, and so has a convergent subsequence. We do a somewhat more complicated thing — we first take a convergent subsequence that is $\ell_{n^1_s}(\varphi_1)$, and then we pick a convergent subsequence only from the elements $\ell_{n^1_s}(\varphi_2)$, and for each index m we choose a sequence of indices for which $\ell_{n^m_s}(\varphi_m)$ is convergent and is a subsequence of the sequences for all smaller indices. Now, each sequence $\{n^m_s\}$ has the property that $\ell_{n^m_s}(\varphi_k)$ is convergent for all $k \leq m$, and the

---

[1] A sequence of linear functionals ℓ_n on a Banach space X is weak*-convergent to ℓ if $\ell_n(x) \to \ell(x)$ for each $x \in X$

sets of indices are decreasing by inclusion. Consider now the sequence $\{n_m^m\}$, i.e. the m-th member of the m-th sequence[2]. This sequence, apart from finitely many elements, is a subsequence of each of our sequences, and so $\ell_{n_m^m}(\varphi_k)$ is convergent for every k. Call the limit $\ell(\varphi_k)$.

If $\|\varphi_k - \varphi_m\|_{C_0} \leq \varepsilon$, then for every n, $|\ell_n(\varphi_k) - \ell_n(\varphi_m)| \leq \varepsilon$, which implies that $|\ell(\varphi_k) - \ell(\varphi_m)| \leq \varepsilon$. This means that ℓ is continuous on the dense subset of C_0 on which it is defined, and thus ℓ can be extended to the whole C_0.

Exercise 10.2. Show that ℓ is a linear functional on C_0 and is positive.

We see that there is a measure μ_j such that for each $\varphi \in C_0(\mathbb{R})$ we have $\int \varphi d\nu_n \rightarrow \int \varphi d\mu_j$. The measure μ_j is a probability measure, as $\mu_j(\mathbb{T}) = \lim \nu_n(\mathbb{T}) = 1$.

The next step after constructing μ_j is to show that it is the measure we want.

For the first if $F_n \rightarrow E$, then $(F_n)_j \rightarrow E_j$.

Exercise 10.3. Show that if $F_n \rightarrow E$, then $(F_n)_j \rightarrow E_j$.

Exercise 10.4. Show that $\operatorname{supp}(\mu_j) \in E$. Hint: Show that for each point not in E there exists a neighborhood for which μ_j vanishes.

Break

To prove that $\int \dfrac{d\mu_j(x)d\mu_j(y)}{|x-y|^{\alpha_j}} \leq K_j$ is not so easy: We can't pass to the limit in y and x separately.

The quickest way to show the estimate is to get outside frames of our course and use the Fourier transform.

Lemma 10.6. *For $\alpha_j \in [0,1)$,*

$$\int \frac{d\nu_n(x)d\nu_n(y)}{|x-y|^{\alpha_j}} = \sum |\widehat{\nu_n}(k)|^2 \varphi(k),$$

---

[2]this is called a diagonal process

where

$$\varphi(k) = \widehat{\frac{1}{|.|^{\alpha_j}}}(k) \geq 0,$$

Proof of the Lemma. We have two claims: First, that $\widehat{\frac{1}{|.|^{\alpha_j}}}$ is positive, and then that $\int \frac{d\nu_n(x)d\nu_n(y)}{|x-y|^{\alpha_j}} = \sum |\widehat{\nu_n}(k)|^2 \varphi(k)$.

We can observe that the function $\frac{1}{|x|^{\alpha_j}}$ is convex and even. Then its Fourier coefficient is $2 \int_0^\pi \frac{\cos(kx)}{|x|_j^\alpha} dx$. If we consider one full period, then the values at the level $\cos(x) = r$ are evaluated at points $s, s+a, t, t+a$ and have the signs $+, -, -, +$. By convexity, the slope (which is the difference divided by a) between s and $s + a$ is greater than the one between t and $t + a$, so the total sum is positive.

Consider $f_m = \frac{1}{|.|^{\alpha_j}} * F_m$, where F_n is a "tent" function. $f_m \to \frac{1}{|.|^{\alpha_j}}$ pointwise.

Exercise 10.5. Show that $\int f_m d\nu_n(x) d\nu_n(y) \to \int \frac{d\nu_n(x)d\nu_n(y)}{|x-y|^{\alpha_j}}$, when $m \to \infty$.

Now $\int f_m d\nu_n(x) d\nu_n(y) = \langle \nu_n, \nu_n * f_m \rangle = \langle \widehat{\nu_n}, (\nu_n * f_m)^\wedge \rangle$
$= \sum |\widehat{\nu_n}(k)|^2 \varphi(k) \widehat{f_m}(k)$. It remains to let $m \to \infty$ and observe that $f_m(k) \to 1$ when $m \to \infty$. $\qquad\square$

As $0 \leq \varphi(k)$ and $\widehat{\nu_n}(k) \to \widehat{\mu}(k)$ we may pass to the limit in $\sum |\widehat{\nu_n}(k)|^2 \varphi(k)$. $\qquad\square$

Now, let us consider $\mathcal{I}_{j,\eta} \subset \mathcal{K}$ such that if $E \in \mathcal{I}_{j,\eta}$ then there exists a finite collection of intervals $\{I_k\}$ which covers E_j and is such that $\sum |I_k|^{\alpha_j + \eta} \leq \eta$.

The good thing about these sets is that $\mathcal{I}_{j,\eta}$ are open and dense in $\mathcal{K}$ which means we can apply the Baire category theorem.

Theorem 10.7 (Baire category theorem[3]). *In a compete metric space an intersection of countably many open everywhere dense sets is non-empty.*

Proof. Let our sets be $\{U_j\}_{j=1}^{\infty}$.

As U_1 is dense there is a point $x_1 \in U_1$.

As U_1 is open there is $r_1 > 0$ such that $B(x_1, r_1) \subset U_1$.

Consider $B_1 = B(x_1, r_1/2)$. As U_2 is dense there is a point $x_2 \in B_1 \cap U_2$.

As U_2 is open there is $r_2 > 0$ such that $B(x_2, r_2) \subset B_1 \cap U_2$.

Take $B_2 = B(x_2, r_2/2)$ and continue the construction in the same manner.

Now, by the construction $\overline{B_j} \subset \overline{B_{j-1}}$ and $\overline{B_j} \subset U_j$.

As an intersection of decreasing sequence of closed sets $\cap \overline{B_j}$ is non-empty. But, by the construction, $\cap \overline{B_j} \subset \cap U_j$. $\qquad\square$

If $E \in \bigcap_n \mathcal{I}_{j,\frac{1}{n}}$ then $\mathcal{H}_{\alpha+\varepsilon}(E_j) = 0$ for each $\varepsilon > 0$, so $\dim_H(E_j) \leq \alpha_j$.

Now if $E \in \bigcap_j \bigcap_n \mathcal{I}_{j,\frac{1}{n}}$ then for every j we have $\dim_H(E_j) \leq \alpha_j$.

But, as $E \in \mathcal{K}$, we know that $\dim_H(E_j) \geq \alpha_j$.

Thus, $\dim_H(E_j) = \alpha_j$ and the construction is exactly as we wanted.

Lemma 10.8. *For every $\eta > 0, j \in \mathbb{N}$ the set $\mathcal{I}_{j,\eta}$ is open (in $\mathcal{K}$).*

Proof. Let $E \in \mathcal{I}_{j,\eta}$. This means that there exists a finite collection of open intervals $\{I_k\}$ which covers E_j. Then $\text{dist}(E_j, \mathbb{T} \setminus \cup I_k) > 0$. Take $0 < \varepsilon < d(E_j, \mathbb{T} \setminus \cup I_k)/j$. Then as soon as $d_H(E, E') < \varepsilon$ we can say that $d_H(E_j, E'_j) < j\varepsilon$, and so the collection $\{I_j\}$ is also a covering for E'_j. $\qquad\square$

The only thing which remains to be proved is that $\mathcal{I}_{j,\eta}$ is dense in K, which we will do the next time.

---

[3]The name of the theorem is connected with (Baire) first and second category sets. A set is called *meagre* if it is closed and nowhere dense (i.e. its complement is everywhere dense). A countable union of meagre sets is called a set of first (Baire) category. All the sets which are not first category are sets of second category. This terminology is a bit odd for topology and must be due to the fact that Baire proved his theorem and introduced the related terms in 1899, when there was no established terminology.

Lecture 11
Hausdorff dimension and sums, part II

The only thing which remains to be proved is that $\mathcal{I}_{j,\eta}$ is dense in $\mathcal{K}$, i.e. that for each $E \in \mathcal{K}$ and each $\epsilon > 0$ there exists $E' \in \mathcal{I}_{j,\eta}$ such that $d_H(E, E') < \epsilon$.

We will actually do this in two step. First we consider a set $\mathcal{R}_\delta$, such that every set R from $\mathcal{R}_\delta$ consists of finitely many of intervals of length at least δ, such that for each j there exists a continuous function g_j supported on R_j such that $\int_{\mathbb{T}} g_j(x)dx = 1$ and $\displaystyle\int_{\mathbb{T}^2} \frac{g_j(x)g_j(y)}{|x-y|^{\alpha_j}}dxdy < K_j$ (notice, that the inequality is strict). It is clear that $\mathcal{R}_\delta \subset \mathcal{K}$. We will prove first that that within δ-distance of each set of $\mathcal{K}$ there exists a set from $\mathcal{R}_\delta$. Then we will prove that within δ-distance of a set from $\mathcal{R}_\delta$ there exists a set from $\mathcal{I}_{j,\eta}$. This will show that $\mathcal{I}_{j,\eta}$ is dense in $\mathcal{K}$.

Lemma 11.1. *Within δ-distance from each set of $\mathcal{K}$ there is a set R which consists of disjoint intervals R_j of length at least δ, and such that for each j there exists a continuous function g_j supported on R_j such that $\int_{\mathbb{T}} g_j(x)dx = 1$ and $\displaystyle\int_{\mathbb{T}^2} \frac{g_j(x)g_j(y)}{|x-y|^{\alpha_j}}dxdy < K_j$.*

Proof. The fact that $E \in \mathcal{K}$ actually means that we have a system $(E, \mu_1, \mu_2, \ldots)$, such that $\text{supp}(\mu_j) \subset E_j$, $\mu_j(\mathbb{T}) = 1$ and $\displaystyle\int_{\mathbb{T}^2} \frac{d\mu_j(x)d\mu_j(y)}{|x-y|^{\alpha_j}} \leq K_j$, which is the same as saying that $\sum |\widehat{\mu_j}(n)|^2 \varphi_j(n) \leq K_j$. Take now $R = E + [-\delta/2, \delta/2]$ and pick a positive test function ψ_δ (i.e. infinitely smooth function with a compact support) such that $\text{supp}(\psi_\delta) \in (-\delta/2, \delta/2)$ and such that $\int \psi_\delta = 1$. Take also $g_j = \mu_j * \psi_\delta$, i.e. $g_j(x) = \int \psi_\delta(x-y)d\mu_j(y)$.

Exercise 11.1. Show that R consists of disjoint closed intervals of length at least δ.

It is easy to see that $\text{supp}(g_j) \subset supp(\mu) + (-\delta/2, \delta/2) \subset E_j + [-\delta/2, \delta/2] \subset E_j + [-j\delta/2, j\delta/2] = R_j$. It is also clear that $\int g_j = \int d\mu_j \cdot \int \psi_\delta = 1$.

If we look at the other way to express $\displaystyle\int_{\mathbb{T}^2} \frac{g_j(x)g_j(y)}{|x-y|^{\alpha_j}} dx dy$, i.e.

$\sum |\widehat{g_j}(n)|^2 \varphi_j(n)$, we see that $\widehat{g_j}(n) = \widehat{\mu}(n)\widehat{\psi_\delta}(n)$. As $\int \psi_\delta = 1$ and $\psi_\delta \geq 0$, one has $|\widehat{\psi_\delta}(n)| \leq 1$. This implies immediately that

$$\int_{\mathbb{T}^2} \frac{g_j(x)g_j(y)}{|x-y|^{\alpha_j}} dx dy = \sum |\widehat{g_j}(n)|^2 \varphi_j(n) \leq \sum |\widehat{\mu_j}(n)|^2 |\widehat{\psi_\delta}(n)|^2 \varphi_j(n) \leq$$

$$\sum |\widehat{\mu_j}(n)|^2 \varphi_j(n) = \int_{\mathbb{T}^2} \frac{d\mu_j(x) d\mu_j(y)}{|x-y|^{\alpha_j}} \leq K_j.$$

We know that $\widetilde{\psi_\delta} \to 0$ and so the second last inequality becomes strict unless μ_j has only finitely many non-zero Fourier coefficients. But if μ_j is a trigonometric polynomial, then it has no more than finitely many zeros, and so its support is $\mathbb{T}$. In this case $E_j = \mathbb{T}$ and we can choose instead $g_j = \frac{1}{2\pi}$ (for which we have $\displaystyle\int_{\mathbb{T}^2} \frac{g_j(x)g_j(y)}{|x-y|^{\alpha_j}} dx dy < K_j$). $\qquad\square$

Thus, R_δ comes within δ of each set in $\mathcal{K}$.

Now it remains to show the following.

Proposition 11.2. *Let a set R consist of disjoint intervals R_j of length at least δ, such that for each j there exists a continuous function g_j supported on R_j such that $\int_{\mathbb{T}} g_j(x) dx = 1$ and $\displaystyle\int_{\mathbb{T}^2} \frac{g_j(x)g_j(y)}{|x-y|^{\alpha_j}} dx dy < K_j$. Then for every $\varepsilon > 0$ there exists $E \in \mathcal{I}_{j^*,\eta}$ such that $d_H(R, E) < \varepsilon$, $E \in \mathcal{K}$ and there exists a finite collection of open intervals $\{I_k\}$ covering E_{j^*} such that $\sum |I_k|^{\alpha+\eta} \leq \eta$.*

78

Proof. Let us start from the observation that for some (possibly large) J we have $R_J = \mathbb{T}$, and so for all $j > J$ we have $R_j = \mathbb{T}$. Whatever were our original g_j we replace them by $g_j = \frac{1}{2\pi}$ for $j \geq J$ (the assumptions of the lemma obviously still hold). Now only first finitely many g_j are non-constant.

Denote $\beta = \alpha_{j^*} + \eta/2$. We may assume that $\alpha_{j^*} + \eta < 1$. We may also assume $j^* < J$.

Now, for a large enough N (how large we will see later) let

$$E = \{\frac{k}{N} : d(k/N, R) \leq \frac{1}{2N}\} + \left[-N^{-\frac{1}{\beta}}, N^{-\frac{1}{\beta}}\right].$$

Exercise 11.2. Show that $d_H(R, E) \leq 1/N$.

Let us first verify the covering condition. It is clear that

$$E_{j^*} \subset \frac{1}{N}\mathbb{Z} \cup \mathbb{T} + \left[-j^* N^{-\frac{1}{\beta}}, j^* N^{-\frac{1}{\beta}}\right],$$

and so is covered by no more than N intervals of length $2j^* N^{-\frac{1}{\beta}}$.
Then

$$\sum |I_k|^{\alpha_{j^*}+\eta} \leq N \cdot (2j^*)^{\alpha_j+\eta} N^{-\frac{\alpha_{j^*}+\eta}{\beta}}$$
$$= N \cdot (2j^*)^{\alpha_j+\eta} N^{(-1-\frac{\eta}{2\beta})} = (2j^*)^{\alpha_j+\eta} N^{-\frac{\eta}{\beta}},$$

which can be arbitrary small when N is large enough.

The tricky part is to show that $E \in \mathcal{K}$.

Choose

$$\Delta_N(g_j(x)) = \begin{cases} \frac{1}{2} N^{\frac{1}{\beta}} \int\limits_{\frac{k}{N}-\frac{1}{2N}}^{\frac{k}{N}+\frac{1}{2N}} g_j(x)dx, & x \in \left[\frac{k}{N} - N^{-\frac{1}{\beta}}, \frac{k}{N} + N^{-\frac{1}{\beta}}\right]; \\ 0, & \text{otherwise.} \end{cases}$$

Let us first look at $j = j^*$.

It is clear that for each k we have

$$\int_{\frac{k}{N}-\frac{1}{2N}}^{\frac{k}{N}+\frac{1}{2N}} g_j(x)dx = \int_{\frac{k}{N}-\frac{1}{2N}}^{\frac{k}{N}+\frac{1}{2N}} \Delta_N(g_j)(x)dx.$$

This implies $\int \Delta_N(g_j) = 1$.

Exercise 11.3. Show that $\mathrm{supp}(\Delta(g_j)) \subset E_j$.

If we show that $\displaystyle\int_{\mathbb{T}^2} \frac{\Delta_N(g_j)(x)\Delta_N(g_j)(y)}{|x-y|^{\alpha_j}}dxdy < K_j$ then we are done.

Now

$$\int_{\mathbb{T}^2} \frac{\Delta_N(g_j)(x)\Delta_N(g_j)(y)}{|x-y|^{\alpha_j}}dxdy$$

$$= \sum_{k,m} \int_{|x-\frac{k}{N}|\leq 1/2N, |y-m/N|\leq 1/2N} \frac{\Delta_N(g_j)(x)\Delta_N(g_j)(y)}{|x-y|^{\alpha_j}}dxdy$$

$$= \sum_{|k-m|\leq s} \int_{|x-\frac{k}{N}|\leq 1/2N, |y-m/N|\leq 1/2N} \frac{\Delta_n(g_j)(x)\Delta_N(g_j)(y)}{|x-y|^{\alpha_j}}dxdy$$

$$+ \sum_{|k-m|>s} \int_{|x-\frac{k}{N}|\leq 1/2N, |y-m/N|\leq 1/2N} \frac{\Delta_N(g_j)(x)\Delta_N(g_j)(y)}{|x-y|^{\alpha_j}}dxdy$$

$$= (I) + (II).$$

(II) If $|k-m| > s$, then

$$\left(1-\frac{2}{s}\right)\left|\frac{k}{N}-\frac{m}{N}\right| \leq |x-y| \leq \left(1+\frac{2}{s}\right)\left|\frac{k}{N}-\frac{m}{N}\right|,$$

so

$$\sum_{|k-m|>s} \int_{|x-\frac{k}{N}|\leq 1/2N, |y-m/N|\leq 1/2N} \frac{\Delta_N(g_j)(x)\Delta_N(g_j)(y)}{|x-y|^{\alpha_j}}dxdy \leq$$

$$\frac{(1+\frac{2}{s})^{\alpha_j}}{(1-\frac{2}{s})^{\alpha_j}} \sum_{|k-m|>s} \int_{|x-\frac{k}{N}|\leq 1/2N,\, |y-m/N|\leq 1/2N} \frac{g_j(x)g_j(y)}{|x-y|^{\alpha_j}} dxdy$$

which is less than K_j when s is suitably large (and the choice of s depends on g_j and K_j, but not on N).

(I) On the other hand, since $\|\Delta_N(g_j)\|_\infty \leq N^{\frac{1}{\beta}-1}\|g_j\|_\infty$, we have

$$\sum_{|k-m|\leq s} \int_{|x-\frac{k}{N}|\leq 1/2N,\, |y-m/N|\leq 1/2N} \frac{\Delta_N(g_j)(x)\Delta_N(g_j)(y)}{|x-y|^{\alpha_j}} dxdy$$

$$\leq \quad (2s+1)N(N^{\frac{1}{\beta}-1}\|g_j\|_\infty)^2 \int_{-N^{-\frac{1}{\beta}}}^{N^{-\frac{1}{\beta}}} \int_{-N^{-\frac{1}{\beta}}}^{N^{-\frac{1}{\beta}}} \frac{1}{|x-y|^{\alpha_j}} dxdy$$

$$\leq \quad (2s+1)N(N^{\frac{1}{\beta}-1}\|g_j\|_\infty)^2 2N^{-\frac{1}{\beta}}\left(\frac{2}{1-\alpha_j}\right)(N^{-\frac{1}{\beta}})^{1-\alpha_j}$$

$$= \quad \frac{4(2s+1)}{1-\alpha_j}\|g_j\|_\infty^2 N^{-1+\alpha_j/\beta} = C_s N^{-\eta/2\beta},$$

so the second term is arbitrary small for sufficiently large N, which proves the estimate for $j=j^*$ (and by the same argument for every $j\leq j^*$).

Exercise 11.4. Prove that

$$\int_{\mathbb{T}^2} \frac{\Delta_N(g_j)(x)\Delta_N(g_j)(y)}{|x-y|^{\alpha_j}} dxdy \to \int_{\mathbb{T}^2} \frac{g_j(x)g_j(y)}{|x-y|^{\alpha_j}} dxdy$$

as $N\to\infty$ when the function g_j is bounded and $j\leq j^*$.

As there are only finitely many $j\leq j^*$ and for each one we have $\int_{\mathbb{T}^2} \frac{g_j(x)g_j(y)}{|x-y|^{\alpha_j}} dxdy < K_j$, we can pick N so large that $\int_{\mathbb{T}^2} \frac{\Delta_N(g_j)(x)\Delta_N(g_j)(y)}{|x-y|^{\alpha_j}} dxdy < K_j$ simultaneously for every $j\leq j^*$.

Unfortunately the estimates don't work for $\alpha_j > \beta$, so we should do something clever.

Break

The construction we have done so far may be still of use, so let us introduce the following notation: we say $\Delta_N(R) = E$, where E is constructed as above and $\Delta_N(g_j)$ is constructed as above. What we have shown so far can be expressed as the following.

Let R be a disjoint union of finitely many closed intervals of length at least δ.

1. $d_H(\Delta_N(R), R) \leq \dfrac{1}{N}$.

2. $\left(\Delta_N(R)\right)_{j*}$ can be covered by finitely many intervals I_k such that
$$\sum |I_k|^{\alpha_{j*}+\eta} \leq (2j^*)^{\alpha_{j*}+\eta} N^{-\frac{\eta}{\beta}} \to 0 \text{ when } N \to \infty.$$

3. If $\operatorname{supp}(g) \subset R_j$ then $\operatorname{supp}(\Delta_N(g)) \subset \Delta_N(R_j) \subset \left(\Delta_N(R)\right)_j$ (the last is when $\frac{1}{N}$ is much less than the length of the intervals composing R.

4. If $g \geq 0$ then $\Delta_N(g) \geq 0$ and $\int \Delta_N(g) = \int g$.

5. If $g \geq 0$ is bounded and measurable, then
$$\int_{\mathbb{T}^2} \frac{\Delta_N(g)(x)\Delta_N(g)(y)}{|x-y|^{\alpha_j}} dxdy \to \int_{\mathbb{T}^2} \frac{g(x)g(y)}{|x-y|^{\alpha_j}} dxdy$$
when $N \to \infty$ for all $\alpha_j \leq \beta$.

Now we do a trick which is not obvious in the circumstances - we increase the variation of the objects.

Let us instead of one set consider a collection of J sets and instead of a sequence of functions $\{h_j\}_{j=1}^{\infty}$ consider a collection indexed by subsets of $[1, J]$.

Let us say that a collection of sets and functions $(\{A_r\}_{r=1}^{J}, \{h_\Lambda\}_{\Lambda \subset [1,J]})$ is "good", if each A_r is a finite union of non-trivial closed intervals, $A_1 + \ldots + A_J = \mathbb{T}$ and

1. If $\#\Lambda > j^*$ then $\sum_{r \in \Lambda} A_r \supset (R)_{\#\Lambda}$, where R is from the assumptions of the Proposition

2. If $j = \#\Lambda \leq j^*$ the following holds.

    (a) $\text{supp}(h_\Lambda) \subset (\bigcup_{r \in \Lambda} A_r)_j$

    (b) $\int h_\Lambda = 1$

    (c) $\displaystyle\int_{\mathbb{T}^2} \frac{h_\Lambda(x)h_\Lambda(y)}{|x-y|_j^\alpha}dxdy < K_j$, for $j \leq j^*$.

Notice that, if we take $A_r = R$ and $h_\Lambda = g_{\#\Lambda}$, then the system obtained is "good".

Let us also say that a system satisfies the η-covering property for $P \subset [1, J]$, $\#P = j^*$ if there is a finite collection of intervals $\{I_k\}$ such that $(\bigcup_{r \in P} A_r)_{j^*} \subset \cup I_k$ and $\sum |I_k|^{\alpha_j + \eta} < \eta / \binom{J}{j^*}$.

Lemma 11.3. *Let us have a "good" system $A = (\{A_r\}_{r=1}^J, \{h_\Lambda\}_{\Lambda \subset [1,J]})$ which has the η-covering property for $P_1, \ldots, P_m \subset [1, J]$. Then for each $P \subset [1, J], \#P = j^*$ and $\gamma > 0$ we can find another "good" system $B = (\{B_r\}_{r=1}^J, \{\widetilde{h_\Lambda}\}_{\Lambda \subset [1,J]})$, such that $d_H(B_r, A_r) < \gamma$ and the system B has the η-covering property for $P_1, \ldots, P_m$ and for P.*

Proof. Observe first that the set of collections A for which we have the η-covering property for P_j is an open set in the product of the Hausdorff metric, and so if γ is sufficiently small then the η-covering property for $P_1, \ldots, P_m$ will be satisfied automatically, so we only need to get it for P.

Now we will take

$$B_r = \begin{cases} A_r + \left[-\dfrac{J}{N}, \dfrac{J}{N}\right], & r \notin P; \\ \Delta_N(A_r), & r \in P. \end{cases}$$

With this choice it is obvious that $d_H(B_r, A_r) \leq \frac{J}{N}$ and is less than each given γ for N sufficiently large.

As $\Delta_N(\bigcup_{r\in\Lambda} A_r) = \bigcup_{r\in\Lambda} \Delta_N(A_r)$ it follows from the properties of Δ_N that for sufficiently large N the new system has the η-covering property for P.

If $\#\Lambda > j^*$ then

$$\sum_{r\in\Lambda} B_r = \sum_{r\in P\cap\Lambda} \Delta_N(A_r) + \left(\sum_{r\in\Lambda\backslash P} A_r + [-J/N, J/N]\right) \supset$$

$$\sum_{r\in P\cap\Lambda} \Delta_N(A_r) + \sum_{r\in\Lambda\backslash P} A_r + [-J/N, J/N] \supset$$

$$\sum_{r\in P\cap\Lambda} (\Delta_N(A_r) + [-1/N, 1/N]) + \sum_{r\in\Lambda\backslash P} A_r \supset \sum_{r\in\Lambda} A_r \supset (R)_{\#\Lambda}.$$

Finally, $B_r \supset \Delta_N(A_r)$. So, for the purpose of dealing with $\#\Lambda \leq j^*$ we may assume that $B_r = \Delta_N(A_r)$ for all r. Again, $\Delta_N(\bigcup_{r\in\Lambda} A_r) = \bigcup_{r\in\Lambda} \Delta_N(A_r)$, so taking $\widetilde{h_\Lambda} = \Delta_N(h_\Lambda)$ we see that for large enough N even the last condition is satisfied for every particular $\Lambda \subset [1, J]$, and as there are only finitely many possibilities we may take N sufficiently large to fit the conditions for all $\Lambda \subset [1, J]$. $\qquad\square$

Back to the proof of the Proposition. We may start from the system where $A_r = R$, $h_\Lambda = g_{\#\Lambda}$ (and there is no collection of indices for which η-covering holds), and by finitely many steps come to a system $(\{E_r\}, \{f_\Lambda\})$ for which the η-covering condition holds for all j^*-element subsets of $[1, J]$. We may also assume that $d_H(R, E_r) < \delta$ in the final collection.

Consider now $E = \cup E_r$. We claim that this is the required set. To begin with, it is obvious that $d_H(E, R) < \delta$. Then, to see if $E \in \mathcal{K}$ we have to consider cases. If $j > J$ we see that $(E)_j \supset E_1 + \ldots + E_J = \mathbb{T}$ and existence of μ_j is trivial. If $j^* < j \leq J$ then $E_j \supset R_j$ and we may pick $\mu_j = g_j(x)dx$. Finally, if $j \leq j^*$ than for arbitrary Λ of cardinality j we have $E_j \supset (\bigcup_{r\in\Lambda} E_r)_j$ and we may take $\mu_j = f_\Lambda(x)dx$.

It remains to check the existence of the covering for E_{j^*}. We see that $E_{j^*} = \bigcup_{\#P=j^*, P\subset[1,J]} (\bigcup_{r\in P} E_r)_{j^*}$, and we can take as a covering the union of coverings for each $(\bigcup_{r\in P} E_r)_{j^*}$. As the system $\{E_r\}$ satisfies the η-covering condition for each P, we see that for the joint covering we have $\sum |I_k|^{\alpha_j+\eta} < \binom{J}{j^*}\eta/\binom{J}{j^*} = \eta$.

Thus, we have found $E \subset \mathcal{I}_{j^*,\eta}$, such that $d_H(E,R) \leq \delta$. $\qquad\square$

Bibliography

[1] Stan Wagon. The Banach-Tarski Paradox. Encyclopedia of Mathematics, Cambridge University Press, 1985.

[2] N. Bourbaki, Theory of Sets. Springer. 1986.

[3] K.J. Falconer, The Geometry of Fractal Sets. Cambridge University Press. 1986.

[4] P.R. Halmos, Measure Theory. Springer. 1978.

[5] Jun Kigami, Analysis on Fractals. Cambridge University Press. 2008.

[6] T.W. Körner. Hausdorff dimension of sums of sets with themselves, Studia Math. 188 (2008), no.3, 287-295.

[7] T.W. Körner. Baire Category, Probabilistic Constructions, and Convolution Square. Lecture notes available at:

https://www.dpmms.cam.ac.uk/~twk/Baire.tex

[8] Pertti Mattila, Geometry of Sets and Measures in Euclidean Spaces. Cambridge University Press. 1999.

Index

μ-measurable functions, 55

Axiom of Choice, 10

Baire category theorem, 75
Banach-Tarski paradox, 4

Cantor sets, 61
cardinal number, 18
cardinality, 18
Continuum hypothesis, 21
covering theorem
    5r, 47
    Besicovich's, 48
    Vitali's, 52

decomposition theorem
    Hahn, 58
    Lebesgue, 60
    Radon-Nikodym, 57

group
    free group, 5

Hausdorff metric, 44, 71
Hausdorff's maximal principle, 11

integral
    Lebesgue, 56
    Riemann, 55
    with respect to a measure, 56

measure
    absolutely continuous, 57

countably additive, 1
finitely additive, 1
Hausdorff-α, 37, 41, 60, 63
inner, 32
Lebesgue, construction, 29, 35
mutually singular measures, 60
outer, 32
Radon, 52
shift-invariant, 3

net limit, 55

ordinal
    closed, 25
    open, 25
    ordinal number, 17, 24

Riesz representation theorem, 63

set
    G_δ, F_σ, etc., 24
    Borel, 23
    Cantor, 37
    closed, 23, 29
    Hausdorff dimension, 66, 71
    Hausdorff-α measurable, 38
    Lebesgue measurable, 35
    Minkowski sum, 71
    open, 23, 29
    self-similar, 43, 63
similitudes
    definition, 43
    open set condition, 63

transitive induction, 51

Well-ordering theorem, 11

Zermelo-Frankel set of axioms, 10
Zorn's lemma, 11

Maria Roginskaya was born in Tula, Russia, and grew up in St. Petersburg (then called Leningrad). In her early years she was involved in training for mathematical competitions. She studied at the St. Petersburg State University. Her research interests span topics in Fourier Analysis, Geometric Measure Theory, and Functional Analysis. For the past 17 years she has held a position at Chalmers University of Technology in Gothenburg. She held also visiting positions at the University of Waterloo (Canada), Université de Lille 1 (France), the Polish Academy of Sciences and the *Fondation Sciences Mathématiques de Paris*.